1007510686

Forging the Methodology that Enlightened Modern Civilization

AMERICAN UNIVERSITY STUDIES

SERIES V
PHILOSOPHY

VOL. 211

PETER LANG
New York • Washington, D.C./Baltimore • Bern
Frankfurt • Berlin • Brussels • Vienna • Oxford

Richard H. Schlagel

Forging the Methodology that Enlightened Modern Civilization

PETER LANG
New York • Washington, D.C./Baltimore • Bern
Frankfurt • Berlin • Brussels • Vienna • Oxford

Library of Congress Cataloging-in-Publication Data
Schlagel, Richard H.
Forging the methodology that enlightened modern civilization / Richard H. Schlagel.
p. cm. — (American university studies. V, Philosophy; vol. 211)
Includes bibliographical references and index.
1. Science—Methodology—History. 2. Science—Philosophy—History.
3. Scientists—History. I. Title.
Q174.8.S334 501—dc22   2011013034
ISBN 978-1-4331-1482-3
ISSN 0739-6392

Bibliographic information published by **Die Deutsche Nationalbibliothek**.
Die Deutsche Nationalbibliothek lists this publication in the "Deutsche
Nationalbibliografie"; detailed bibliographic data is available
on the Internet at http://dnb.d-nb.de/.

Cover illustration by Edouard Riou (1870) of Galileo demonstrating his
telescope in Venice in 1609.

The paper in this book meets the guidelines for permanence and durability
of the Committee on Production Guidelines for Book Longevity
of the Council of Library Resources.

© 2011 Peter Lang Publishing, Inc., New York
29 Broadway, 18th floor, New York, NY 10006
www.peterlang.com

All rights reserved.
Reprint or reproduction, even partially, in all forms such as microfilm,
xerography, microfiche, microcard, and offset strictly prohibited.

Printed in Germany

A good scientist should never be so arrogant as to be *certain* about anything. Never, that is, apart from on one point: that what we refer to as the modern scientific method is *non-negotiable* in its all-encompassing importance as a worldview. Many would argue it is the *only* worldview that a rational thinking person can have in explaining how and why the world is the way it is.

Jim Al-Khalili
Pathfinders: The Golden Age of Arabic Science

Table of Contents

Introduction ix

Chapter 1: Our Greek Heritage Including Plotinus..................................... 1
Chapter 2: Decline of Scientific Inquiry in the Medieval
           Period and the Arabic Revival..................................... 11
Chapter 3: The Early Scholastics and Their Revisions of Aristotle's Philosophy 19
Chapter 4: Contributions of Grosseteste and Later Scholastics 25
Chapter 5: Copernicus, Kepler, and Gilbert 33
Chapter 6: Galileo 51
Chapter 7: The Divergent Philosophies of Bacon and Descartes 81
Chapter 8: Newton's Incomparable Achievements 109

Epilogue... 139
End Notes 151
Index.. 157

Introduction

The influence of science and technology is so prevalent in western societies today that it is easy to take for granted how much the standards of living, advancements of knowledge, medical developments, educational opportunities, liberating and equitable social and political institutions, along with world travel and nearly instantaneous communication and dissemination of information are owed to these achievements. Yet visiting those societies that have not yet acquired the benefits of western science and culture, as in most countries of Africa and some in the Middle East and South America, and to a lesser degree now in regions of China and India, offers a striking reminder of the difference.

Moreover, those who disparage this progress owing to the greater devastation during the First and Second World Wars due to the advances in technology that produced gas warfare, more lethal artillery, horrendous bombing raids, V-1 and V-2 rockets, and the incredible radiational incineration spewed by the atomic bomb, overlook that it was not the advanced weaponry that was at fault, but fanatics like Hitler who incited the wars. While Einstein's formula $E = mc^2$ and Lise Meitner's theory of nuclear fission, along with the effort of the most brilliant team of theoretical physicists and engineers ever assembled, led to the success of the Manhattan Project and the detonation of two atomic bombs, it was the heads of state who decided how it would be used. The realization that Werner Heisenberg was directing nuclear research in Germany with the intention of creating an atomic bomb led to Niels Bohr informing Churchill and Leó Szilárd and Einstein advising President Roosevelt that their countries must create the weapon before Hitler, otherwise the Third Reich would win the war and dominate the world.

Furthermore, the decision to drop the bomb on Hiroshima and then on Nagasaki was not made by the scientists that created the bomb, such as Enrico Fermi, James Franck, and J. Robert Oppenheimer who were opposed to using it, but by President Truman. It was his decision that dropping the bombs was necessary to force the Japanese to surrender without having to invade Japan that could have resulted in many more casualties. Later, during the cold war between the Americans and the Russians, long range warfare was made possible by the development of intercontinental nuclear warheads or missiles leading to the Cuban missile crisis that was resolved owing to the threat of "mutual destruction."

Again, however, it was the heads of state and generals who were responsible for making the decisions and providing the financing to create the advanced military technology to serve their national interests. Similar arguments can be offered to refute the charges that technological developments also were responsible for the dreadful initial exploitation of unskilled workers during the industrial revolution, creation of modern urban ghettos, environmental degradation, and climate change, but that would be like blaming the Christian religion for the current pedophiliac scandal of the Roman Catholic Church or Islam for the worldwide Jihads, rather than the clerics and terrorists responsible for these deviant acts.

Instead of blaming the institutions for these depravities, they should be attributed to the pervasive weaknesses of human nature that produced them: irrationality, avarice, egocentrism, aggression, sadism, and the lust for and fixation on power. It no longer being credible to ascribe these tragic human failings to original sin, they now can be attributed to our evolutionary heritage driven by competitive natural selection or "survival of the fittest" as encoded in our genes. As with social conventions, political institutions, legal structures, economic systems, and personal relations, the effective utilization of scientific research and technology can only be as humane or enlightened as their use by human beings.

Another reason there seems to be such an unappreciative attitude toward contemporary science is that unlike the Enlightenment when savants were extolling the accomplishments and promise of modern science in contrast to the stifling feudal system it was replacing, most people today ignore the horrific conditions that prevailed throughout history: the poverty, disease, ignorance, illiteracy, and natural disasters, along with the universal exploitation and repression by the prevalent tyrants or autocrats.

A further cause of the aversion to science is that its resultant worldview has discredited the supernatural framework of the world's religion's that provided so much spiritual comfort, support, and moral direction in the past, as it still does in the present. Dislodging human beings from the center of the universe and explaining their origin as due to natural conditions rather than a special creation, replacing miracles by scientific explanations and eliminating the credibility of such Christian doctrines as the virginity of Mary and the virgin birth of Jesus (who would have lacked the male chromosomes for a normal birth), and understanding the molecular impossibility of the transubstantiation of the Eucharist has eroded the belief in Christianity for those who comprehend the significance of these developments.

Moreover, having discovered that mystical experiences, feelings of blissfulness, and hearing divine commands are due to localized neurophysiological processes and excluding any transcendent meaning to human existence given the distressing state of the universe and human existence as we know it, science is looked upon as the destroyer of cherished, consoling beliefs, rather than as the liberator from ignorance and superstition and ameliorator of the human condition.

It is the hope of the author that when confronted by the remarkable discoveries, theoretical explanations, cognitive transformations, and technological advances that brought about these developments the reader will be better able to accept scientific inquiry as the intellectual and social liberator that it has been, as well as the most effective means for improving the dreadful living conditions of the past.

But my focus will be on how this understanding has been acquired, rather than on the changes that it has brought about. This is because the former conforms more to my research and publishing endeavors and also because the other aspect has been excellently described by Timothy Ferris in his recent book, *The Science of Liberty*,[1] that I commend to anyone interested in knowing what a difference scientific inquiry has made to the advent of modern civilization.

CHAPTER ONE

Our Greek Heritage Including Plotinus

The emergence of an empirical-rationalistic orientation eventually dispelling the previous mythopoetic rendering of the world by a more realistic explanation was initiated by the ancient Greeks. Their legacy from the seventh century BC first includes the Milesians—Thales, Anaximander, and Anaximines—whose cosmological explanations of the universe, as arising respectively from Water, the Unbounded, and an Air-Substrate, were more naturalistic than a divine creation, as was Anaximander's account of land animals evolving from aquatic creatures. There followed Anaxagoras' principle that since "being cannot come from not-being nor perish" what exists must be "ultimate and indestructible;" Empedocles' doctrine of the four elements, "fire, air, earth, and water;" and Leucippus and Democrates' prescient atomic theory. Philolaus' attribution of "an oblique circular motion" to the earth around a central fire, along with Heraclides of Pontus' addition of a rotational motion from west to east (with the stars fixed) to account for the apparent rising and setting of the sun and determination that Mercury and Venus revolve around the sun, led to Aristarchus' heliocentric theory. Yet it was Ptolemy's geocentric astronomical system that prevailed until Copernicus adopted heliocentrism owing to its greater harmony and mathematical simplicity, citing the Pythagoreans as his predecessors.

Along with these developments there was Hippocrates' famous medical school at Cos and the physiological investigations of Herophilus and Erasistratus, as well as the impressive mathematical discoveries of the Greeks beginning with the

Pythagoreans who, according to Aristotle, "regarded numbers as the elements of all things, and the whole heaven as a musical and numerical scale;" Hippocrates of Chios whose work compiling the *Elements of Geometry* preceded Euclid; Eudoxus of Cnidus who made the incredibly accurate determination of the solar year to be 365 days and 5 hours and who originated the view that the celestial bodies revolve on a series of concentric spheres with the earth in the center; and Archytas of Tarentum who maintained that the universe was infinite and eternal and solved the famous problem of "the duplication of the cube." These were remarkable achievements especially when contrasted with the mythical or mystical accounts of the various religions, such as the Old Testament Genesis depiction.

In addition, there were the tremendous contributions of Plato and Aristotle, the formers' marvelous dialogues and profoundly influential metaphysical system composed of three eternal components: the Receptacle or spatial-temporal locus of created entities in the empirical world; the transcendent existence of the Realm of Forms or ideal archetypes culminating in the Good that were the referents of all true knowledge and moral principles, as well as the model of creation imperfectly exemplified in the Receptacle; and the Demiurge whose function it was to impose as much as possible the Ideal Forms on the imperfect, intractable contents of the Receptacle leading Plato to refer to any account of the physical world as "a likely story." In addition to its influence on mathematicians because of its being the method of achieving knowledge of the Realm of Forms, it also inspired Plotinus' Neoplatonic conception of Divine Emanations.

Then there was the tremendous influence of his pupil Aristotle who, after studying in Plato's Academy for twenty years, produced the extensive treatises on physics, astronomy, metaphysics, logic, categories, the soul, dreams, biology, generation and history of animals, ethics, and politics that tended to dominated Western thought, as interpreted by the scholastics, from the 13th to the 17th centuries. Often regarded as the "Authority" of all knowledge during that period, he could justly be considered the most comprehensive thinker of all time.

With his death in 322 BC there followed the Hellenistic Period during which the center of learning shifted from the renown Academy of Plato and Aristotle's Lyceum in Athens to the famous Library and Museum established by the Greek Ptolemies in Alexandria that was founded by Aristotle's pupil Alexander the Great in 332 BC. As Charles Singer states:

> From 300 B.C. to A.D. 200 most eminent men of science were teachers at Alexandria. A few, notably Archimedes and Galen, were less intimately linked with the Egyptian metropolis. Yet even they were pupils and corresponded with Alexandrian teachers. Greek science from about 300 BC onward is thus not inadequately described as "Alexandrian Science."[2]

Among its outstanding early scholars were Eratosthenes the librarian who devised an ingenious method for measuring the circumference of the earth, invented the famous "Sieve" for investigating prime numbers and also measured the obliquity of the ecliptic; Euclid, whose early schooling had been in Plato's Academy, was among the first to be attracted to Alexandria where he wrote his *Elements of Geometry*, parts of which are included in geometry books today and which has been described as "the greatest textbook ever written;" Apollonius of Perge, who is reported to have "spent a very long time with the pupils of Euclid at Alexandria," is famous for his investigation of conic sections and introduction of the terms 'ellipse,' 'parabola,' and 'hyperbola' to designate the various curves produced by three different angular sections of the cone which were instrumental in Kepler's discovery of his three laws of planetary motion; Hipparchus who, owing to his numerous astronomical investigations, detected the precession of the equinoxes and introduced "epicycles" and "eccentrics" to explain the observed variations in size and brightness of the planets that were inconsistent with Eudoxus' circular orbits; Strabo the geographer and cartographer; Galen whose accurate anatomical and physiological studies based on dissections of dead and living animals enabled him to construct an ingenious physiological system remarkably accurate for the period; Hero of Alexandria whose playful contrivances and clever practical instruments enhanced his investigations in optics and dioptrics and whose *Mechanics* shows "understanding of the cogwheel, of rack and pinion, of multiple pulleys, of transmission of force from a rotating screw to an axis at right angles to it, and to the combination of all these devices with levers" (p. 86), and Ptolemy who created important astronomical instruments and whose *Almagest* employing the epicycles and eccentrics of Hipparchus to explain the (illusory) retrograde motions of the superior planets and the variations in brightness and distance of each of the planets replaced all previous astronomical systems until Copernicus's *De Revolutionibus orbium coelestium* published in 1543.

Studying in Sicily rather than Athens or Alexandria but corresponding with the scholars at Alexandria, Archimedes (287-212 BC), renown for his outstanding contributions to mathematics and physics, was also famous for his ingenious technological instruments, such as levers and pulleys, used in the defense of Syracuse during the siege of the Romans. He invented the mechanical screw to raise water and experimentally proved the principle of specific gravity, evidence that the Greeks were not remiss in performing rudimentary experiments to develop and test their theories, as was true of Galen, Hero, and even Aristotle whose investigations dissecting insects and animals were highly praised by Darwin. Thus they are rightly credited with initiating research in most branches of science, including the first attempts to explain the origin of the universe naturalistically, pursuing empirical medical investigations, introducing such formalisms as deductive logic, geometry, and the methods of exhaustion and integration that were precursors of calculus, inventing numerous technological instruments and devices that constituted the foundations and inspiration for the later developments of modern classical science.

Before leaving this discussion of the marvelous mathematical and scientific legacy of the ancient Greeks, some mention should be made of Plotinus whose philosophy of Neoplatonism has been described as "the culmination of all Greek philosophy"[3] (although Clark's claim is hardly justified since his philosophy is completely devoid of any scientific content). He lived from 205-270 AD and though he made no contributions to mathematics or science, his orientation being completely mystical, the considerable influence of his theory of emanation on some Muslim scientists and the scholastics justifies some description of his philosophy.

A native of Egypt but perhaps of Roman descent, at age twenty-eight he went to Alexandria where he became a disciple of Ammonius Saccas for ten years and then formed an association with Longinus and Origen who were among the greatest of the Greek Fathers of the early church. After a distinguished career in the service of several Emperors, at the age of forty he established a society of philosophers in Rome where he wrote his great work, the *Ennaeds,* and taught until his death. Owing to his weak eyesight it was Porphyry, one of his students, that he entrusted with the task of editing his manuscript and who wrote the first biography of him. Because Porphyry arranged the manuscript into six books with each book containing nine Tractates (treatises), it acquired the name *Ennaeds,* meaning nine treatises. As I wrote previously, "Porphyry published the *Ennaeds* at the beginning of the fourth century when Christianity was about to be proclaimed the official religion of the Empire by Constantine. Although Neoplatonism was a pagan philosophy, its devout mystical content made it congenial to the emerging Christian theology that was to dominate and transform philosophical inquiry for the next thousand years."[4]

According to Porphyry, Plotinus described four mystical visions that were the basis of his reinterpretation of Plato's tripartite philosophy as three superimposed domains called "*hypostases,*" variously translated as the "One" or the "Supreme," the "Intelligible World" or the "Divine Mind," and the "World Soul" or the "Universal Soul." (p. 417) The last two hypostases, the "Intelligible World" and the "World Soul" are further divided, the former containing "Intelligible Beings" and the latter animals, disembodied souls and demons, along with the souls embedded in humans and plants.

His conception of the manner of creation as an "emanation," "overflowing," or "radiation," though influenced by Plato, is exceedingly original derived from his mystical visions. Beginning with the "One" which is beyond Being, each *hypostasis* owing to its "abundance" or "effulgence" emanates this superfluity of reality producing the next successive *hypostasis*. Called an "apostasy" or "falling away," these lesser emanations are complemented by each *hypostasis* "aspiring" to return to its source. Exemplifying the tremendous influence of Plato, the "One" was modeled after his conception of the "Good," the "Intelligible World" on his theory of the Realm of Forms presented in the *Republic*, with the "World Soul" inspired by his notion of the Soul in the later *Timaeus*.

This conception of the three *hypostases* is more precisely described in the Second Tractate of Book V. Beginning with the "One" (which perhaps was influenced by Parmenides' "One" or "Unit"), he states that as the source of Being it is "above Being, without limits, division, plurality, or qualification, hence a Supreme Unity."

> It is precisely because there is nothing within the One that all things are from it: in order that Being may be brought about, the source must be no being but Being's generator, in what is to be thought of as the primal act of generation. Seeking nothing, possessing nothing, lacking nothing, the One is perfect and, in our metaphor, has overflowed, and its exuberance has produced the new: this product has turned again to its begetter and been filled and has become its contemplator and so an Intellectual Principle.[5]

Based on his mystical visions these concepts are obscure but presumably intelligible to their author. As Plotinus admits, the mystical nature of this experience "is why the vision baffles telling; we cannot detach the Supreme to state it [...] which is to be known only as one with ourselves." (VI, 9:10, p. 624) Transcending the intellect, the One is ineffable, "describable negatively as not limited, pluralistic, qualitative, or at rest or in motion." "Only by an ascending process of purification and disembodiment of the soul [can] one attain both self-knowledge and contemplation of the One. At its purest the soul finds itself in the "presence" of the One without being submerged in it or losing its self-identity.... (p. 418)

As for the second *hypostasis,* the World Intellect or Divine Mind, Plotinus synthesizes the views of Plato and Aristotle, rejecting Plato's theory of the Forms as existing independently of the Demiurge, but like Aristotle's conception of the Prime Mover who thinks on himself, he assigns them to the Divine Mind as its objectified thoughts. Only in the realm of lesser souls is there a separation between the mind and its contents. In contrast to the One, the Divine Mind can be considered a "Real Being" that contains the Forms within itself as a unified mind, while all existing things are its "images" or projections similar to Plato's Receptacle.

The third *hypostasis*, the "World Soul" or the "Universal Soul" comprising the lower Cosmic World, is instantiated as "daemons" translated as "Supernals," "Celestials," or "Divine Spirits" that are spiritual powers below the Gods but above man. Other instantiations take the form of the embodied souls in men which are called "amphibians" because they also aspire to return to the higher Celestial Realm, along with inhering in animal and vegetal forms and "emanating" material bodies. These amphibious souls are distinguished from the unembodied souls that exist in the realm of the World Intellect. Though non-spatial, the World Soul individuates into particular "forms," "bodies," or "masses" while remaining in "sympathetic unity" with other souls. (Cf. p. 419) How Singer could conceive this worldview as the "culmination of Greek philosophy" is beyond comprehension, especially for such a fine scholar.

Plotinus describes the soul in man as having three functional components: the "Intellectual, the Reasoning, and the Unreasoning." In its highest actualization in the purified person the Intellectual Soul intuits the World Intellect while also aspiring to contemplate the One. This individual "has attained the mystical vision, has reached a God-like epistemic state analogous to Plato's *noesis* or intelligence." (pp. 419–420) The Reasoning Soul or discursive intellect, in contrast, has acquired wisdom by dialectical reasoning but not attained divine illumination. The Unreasoning Soul is the next lowest which exists in animate beings and in humans as the source of feeling and sensing. But the World Soul has an additional creative power that is the source of corporeality, an imperceptible kind of Non-Being which yet exists.

Resembling Plato's Receptacle and Aristotle's prime matter, as the three dimensional domain of the Cosmic World it is the most extended emanation devoid of all qualities, yet is the source of evil. The latter, the most difficult to explain in any religious worldview, is not considered a positive factor but the result of an expended goodness when the Divine emanation reaches its weakest limits. This unformed matter is ineffable like the One, but resembles darkness while the One is likened to a "luminous radiation." It is the latter which is the true home of the human soul which can be reached by denouncing all desires and passions, thus becoming "disengaged from the body thereby attaining a state of purification" which, he says, "would not be wrong to call "Likeness to God" or to the "Supreme Exemplar." (I, 2:7; p. 36)

While the authenticity of Plotinus' experiences for him cannot be questioned, in terms of what we know today there can be little doubt that they were figments or projections of his brain, rather than descriptions of a spiritual reality. His own account indicates that they were what are now called out-of-the-body experiences. "Many times it has happened: lifted out of the body into myself; becoming external to all other things and self-encentered; beholding a marvelous beauty; then, more than ever, assured of community with the loftiest order; enacting the noblest life, acquiring identify with the divine [....]. (IV, 8:1; p. 357)

As I maintained in *The Vanquished Gods*, the experiences of the founders of Christianity and other religions, as reported in their sacred writings, included revelations, dreams, ecstatic visions, inner voices, states of bliss, and what are now known as out-of-the-body experiences or astral projections, and near death encounters.[6] Having no alternative explanation of their occurrence, they were accepted on face value as evidence of another reality, along with the existence of angels, demons, devils, talking serpents, and spirits, such as Gabriel and the Holy Spirit. Even now evangelicals argue that mystical experiences and revelations are evidence of another dimension of reality complementing the physical world and thus cannot be explained scientifically.

However, studies of abnormal behavior, brain damage, post-traumatic stress disorders, the effects of mind-altering drugs or herbs, and advanced knowledge of the neurophysiological functions of the brain provide compelling evidence that these seemingly transcendental experiences, along with terrifying dreams, realistic hallucinations, and delusional states, can be localized in and explained as aberrations of the brain. Epileptics experiencing temporal lobe seizures, psychotics with temporal lobe disorders, migraine sufferers, schizophrenics, as well as those who have taken mind altering drugs or herbs report similar traumas. Michael A. Persinger claims, based on on clinical research, that "God experiences are the products of the human brain...tempered by the person's learned history...."[7]

> God Experiences are predicted to be correlated with transient electrical instabilities within the temporal lobe of the human brain. These temporal lobe transients (TLTs) are normal changes that are precipitated by maturation, personal dilemma, grief, fatigue, and a variety of physiological conditions. Productions of TLTs create an intense sense of meaningfulness, profundity, and conviction. (p. x)

E. Slater and A. W. Beard report similar bizarre symptoms of psychotics who also suffer from temporal lobe aberrations.

> Mystical delusional experiences are remarkably common. One patient said that God, or an electrical power was making him do things; that he was the Son of God. Another said he felt God working a miracle on him. Another felt that God and the Devil were fighting within him and that God was winning.... Hallucinations were often extremely complex and were usually full of meaning, often of a mystical type. Nearly always there were auditory hallucinations at the same time. One patient saw God, heard voices...and received a message that he was going to heaven. Another had a vision of Christ on the cross in the sky [analogous to Constantine's vision preceding the battle at the Mulvian bridge], and heard ...God saying: 'You will be healed, your tears have been seen.'[8]

As I reported in my book, *The Vanquished Gods*, Kenneth Dewhust and A. W. Beard reviewed the conversion experiences of such mystics and saints as

> St. Paul, St. Teresa of Avila, Joseph Smith (founder of the Mormon religion), Francis Liebermann (a Jewish convert to Christianity), Hieronymus Jaegen (a German mystic), St. Thérèse of Lisieux, and Catherine dei Ricci (a Florentine saint), all of whom exhibited some typical symptoms of temporal lobe epilepsy... terrifying visual and auditory hallucinations, visions of a pillar of light, temporary loss of consciousness or sight, mystical unions, and cosmic visions. (p. 306)

Furthermore, the history of religions are replete with accounts of ascetics, monks, hermits, and shamans retreating to caves or isolated cells where, by intense meditation, fasting, or other sensory depravation conditions, they induced the kinds of bizarre subjective experiences mentioned previously. But the most extraordinary were those described by Muhammad. When he was approaching forty years of age he became obsessed with religious questions, often withdrawing during the holy month of Ramadan to a cave at the foot of Mt. Hira three miles from Mecca where he spent many days and nights in fasting, meditation, and prayer. One night in the year 610 AD, as he was alone in the cave, the pivotal experience of all Mohammedan history came to him.

> As described by Muhammad ibn Ishaq his major biographer: "Whilst I was sleep, with a coverlet of silk brocade whereon was some writing, the angel Gabriel appeared to me and said, 'Read!' I said, 'I do not read.' He pressed me with the coverlets so tightly that Methought 'twas death. Then he let me go, and said 'Read!'.... So I read aloud, and he departed from me at last. *And I awoke from my sleep*, and it was as though these words were written on my heart. I went forth until...I heard a voice from heaven saying, 'O Mohammed! thou art the messenger of Allah, and I am Gabriel.' I raised my head toward heaven to see, and lo, Gabriel in the form of a man, with feet set evenly on the rim of the sky, saying 'O Mohammed! thou art art the messenger of Allah, and I am Gabriel.'"[9]

This certainly is a most extraordinary account. Because he was illiterate, when he awoke from these dreams he dictated these revelations to an amanuensis over a number of decades. It is said that "[o]ften, when they came he fell to the ground in a convulsion or swoon; perspiration covered his brow." When asked to describe how he received the revelation, according to Will Durant he replied

> "that the entire text of the Koran existed in heaven, and that one fragment at a time was communicated to him, usually by Gabriel. Asked how he could remember these divine discourses, he explained that the archangel made him repeat every word. Others who were near the Prophet at the time neither saw nor heard the angel."[10]

Though this discussion of Plotinus' mystical system and analysis of the neurophysiological basis of religious experiences was a deviation from the review of Greek philosophy, I believe it was relevant because, in addition to preparing the reader to understand later references to Neoplatonism, it illustrated the incongruous nature of mystical experiences and the irreconcilable differences between the scientific and religious worldviews despite the attempts by some religionists to reconcile them. It also should help explain why the dominance of Christianity for over a millennium precluded any scientific inquiry and why the *radical* defenders of the Muslim religion today, who feel threatened by western civilization, are attempting to impose on other Muslims their medieval worldview and Sharia, Islamic law based on the Koran.

The detriment of this worldview to the intellectual, cultural, technological, and economic development of the Muslim civilization throughout the Middle East is obvious, except in the United Arab Emirates which has been open to western influence, especially its technology. But now even Saudi Arabia is committing billions of dollars over the next few years "to convert a stretch of desert into a buzzing hub of scientific research and development, with cutting edge universities...."[11] Yet the outrageous fraudulent Iranian election, confirmed in Ayatollah Khameini's declaration that "the people have voted but God has decided," to preserve its Muslim autocracy in opposition to overwhelming public support for a freer, more liberal democratic government, is a chilling reminder of how theocracies must resort to blatant lies, false accusations (claiming that the British and American governments were the main instigators of the demonstrations), and brutal force to maintain their control. Its leaders having to rely on such immoral, unjust, and cruel methods reveals the dark side of religions—as also typified by the Catholic Church during the Middle Ages—despite their profession of spiritual values.

But as I write this, the present successful antigovernment revolution of millions of courageous, determined Egyptians from all classes of society have finally succeeded in forcing President Hosni Mubarak to relinquish his three decades of corrupt, repressive, autocratic control of the country in the hope of creating a democratic, republican government. Initiated by a similar revolt in Tunisia, this desire for greater liberty and political representation seems to be sweeping the Arabic world.

CHAPTER TWO

Decline of Scientific Inquiry in the Medieval Period and the Arabic Revival

The extraordinary intellectual tradition initiated by the ancient Geeks was suspended by the dominance of the Roman Empire whose unparalleled conquests, monumental engineering feats displayed in their splendid aqueducts and temples, beautiful mosaics and statues, and marvelous literary creations did not extend to producing a single outstanding scientist or mathematician who was not Greco-Roman. Then with Emperor Constantine's adoption of Christianity as the official religion of the Roman Empire in 313 AD and transfer of its seat to Constantinople, the supernatural worldview of Christianity further obviated any pursuit of science or mathematics.

As Saint Augustine, Bishop of Hippo, who lived in the fourth century dogmatically stated: "Nothing is to be accepted except on the authority of Scripture, since greater is that authority than all powers of the human mind,"[12] a religious orientation that cast the long shadow on human history known as the Dark Ages or Medieval Period. Since early Christians believed that the world would end with the last judgment determining whether one would attain eternal salvation or go to purgatory, their primary aim was to gain salvation after death and thus they were indifferent to the kinds of empirical inquiries or theoretical questions pursued by the ancient Greeks, as indicated in the following statement by Bishop Ambrose, one of the Patristic Fathers:

"To discuss the nature and position of the earth does not help us in our hope of the life to come. It is enough to know what Scripture states, 'that He hung up the earth upon nothing.' (Job, xxvi, 7) Why then argue whether He hung it up in air or upon the water [the views respectively of Anaximines and Thales].... Not because the earth is in the middle, as if suspended on even balance [the position of Anaximander], but because the majesty of God constrains it by the law of His will, does it endure stable upon the unstable and the void."13 (Brackets added)

One hardly could find a better example of how using God to explain things terminates any empirical inquiry. This was especially true during the later Middle Ages when the Catholic Church was dominant, one of the most degrading periods in human history. The successful growth of Christianity following the proselytizing journeys of Saint Paul was largely due to Jesus' teachings extolling the "brotherhood of man," urging that "one should do unto others as one does unto oneself," along with treating the "humble, meek, and downtrodden" with compassion as the chosen people of God—rather than the rich and the powerful. This was an extraordinary message for the time and had a forceful appeal among some Romans who found their lives bleak.

But by the late Middle Ages the Catholic Church, which claimed the exclusive mantle of Christianity due to its legacy with Jesus' disciple Peter, had acquired vast properties, enormous wealth, built splendid cathedrals, and exerted tremendous power over their extensive domains throughout Europe. Contrary to the teachings of Jesus, its domination typically was repressive, threatening, and extortionist. The pope insisted on incontrovertible authority over the interpretation of Christian doctrine and observance of the sacraments reinforced by preaching in Latin to the illiterate masses so that the laity were unable to question its authority. Those like John Wycliffe who wanted to translate the scriptures into vernacular English to make them accessible to the few who could read were charged with heresy, threatened with excommunication and imprisonment. Maintaining an aura of mystery, infallibility, and the spiritual transcendence of its clergy, the Catholic Church adamantly refused any mediation or intersession, other than their own, between the laity and God. This was to prevent the laity from attempting to communicate directly with the deity, thereby insuring their absolute dependence on the Church's blessing for salvation. Martin Luther will address this along with other church abuses, such as absolutions, in his famous Ninety-five Theses against the papacy printed in 1517 to reform the Catholic Church.

As if that were not bad enough, Jesus' emphasis on brotherly love, charity, and humbleness was generally disdained by the church hierarchy. The Church (along with the nobility and aristocracy) owned nearly all the peasants' land and exacted so much rent the latter commonly were forced to lived in near destitution. Thousands died from bad harvests or plagues with little or no support from the Church. Furthermore, the Popes, Cardinals, Archbishops, and Bishops were constantly demanding donations for indulgences, absolutions, sacraments, and tithes to support their opulent life styles, church constructions, and incessant wars. Although nearly impoverished, the peasantry usually was kept from rebellion by the threat of excommunication and fear of being damned and sent to purgatory for their sins, lacking the intersession of the church. One can only cringe when imagining how Jesus would have regarded this considering his angry overturning of the tables of the money changers in the temple.

In striking contrast to the abysmal condition of the peasantry, the *higher* clergy enjoyed great wealth living in luxury, opulence, and security, along with dining in splendor. One reads of the golden goblets, silver service, fine porcelain, magnificent tapestries, paintings, and sculpture adorning their palaces (as can be seen today in the Vatican and its museums). The higher clergy wore luxurious silk robes trimmed in ermine with silver and gold jewelry and holy relics topped by ornate miters and carried golden scepters to exalt their status (as also seen today). The vow of chastity did not deter a number of clergy from having mistresses and illegitimate children while some monasteries and convents had the reputation of being cloisters of iniquity.

But even more abhorrent than those abuses was the manner in which the Church maintained its control. The laity lived under the constant threat of being charged with heresy for the slightest criticism of Catholic doctrine or resistance to the extortions of the clergy, along with any infraction of Church sacraments or ritual. This could result in secluded imprisonment in a dark, damp dungeon sleeping on the cold stone floor, served food hardly fit to eat with turbid water, smelling one's own excrement, and watching rats scampering about. As in ancient Rome, canon law required two witnesses to bring a charge of heresy, but given the authority and threatening powers of the Church this was not difficult to obtain. Still, the Church did sponsor considerable charitable work and administer the required sacramental duties, such as baptism and last rites, but these were performed mainly by the parish priests who were of a much lower class and often overworked, illiterate, and impoverished, not the church hierarchy.

I realize this is a very severe indictment with little exceptions or qualifications which of course there were, but I believe that as a generalization it is supported by historical evidence.[14] Even the fact that the present Pope Benedict XVI and archbishops have resisted addressing the widespread pedophiliac scandal is further evidence. When describing this one is reminded of the civil conditions in Russia under the communist rule of Stalin. There too official doctrine, in that case dialectical materialism, was the sacred creed not to be questioned or criticized without risking being sent to a gulag or Siberia, thus consigned to horrible living and working conditions, sometimes tortured or made to "disappear." Anyone could make the slightest accusation of another whom he disliked or wanted removed without recourse to civil rights or legal defense.

In his book mentioned previously, Timothy Ferris documents how authoritarian regimes, such as Hitler's National Socialism, Stalin's Soviet Union, Mussolini's Fascism, and the appropriately named medi*eval* Catholic Church (though probably not as extensively or extremely), because of their claim to the absolute truth were intolerant of criticism and free speech, stifled opposition, rejected democratic processes and individual rights, along with stifling scientific inquiry because of their claim to a higher truth. This is why the democratic countries with their freedom of inquiry have been able to prevail over the repressive autocracies.

Turning to a different crucial development during the later Middle Ages, it was the Arabic Muslims who revived the inquiries of the Ancient Greeks owing to the preservation, translation, and emendation of their treatises from about the ninth to the twelfth centuries. Unlike the early Christian Church fathers, Muhammad maintained that because scientific research revealed the majesty of Allah's creation of the world it also could augment veneration of Him. Thus he "besought his disciples to seek knowledge from the cradle to the grave…for 'he who travels in search of knowledge, travels along Allah's path to paradise'."[15] Then, owing to the extensive conquests of the Muslims with the founding of Baghdad in 750 by the Abbasid Caliph al-Mansur and the contributions of later Caliphs, such as Harun al-Rashid and particularly al-Mamun who built the Baghdad Academy of Science with its library and observatory, called the "House of Wisdom," the Abbasid capital gradually became the center of learning replacing Athens and Alexandria as the apex of scientific and mathematical studies. As Majid Fakhry states: "it was during al-Mamun's reign in the ninth century that the translation of medical, scientific and philosophical texts, chiefly from Greek or Syria, was placed on an official footing."[16]

Once the translations were available, Arabic speaking scholars for the following five centuries wrote commentaries on the Greek treatises that emended and extended their original research.[17] In the space available only a few of the major contributors can be mentioned. The names of al-Kindi and al-Khwarizmi stand out among those of the first half of the ninth century. Referred to as the Arabic philosopher king for initiating philosophical research in Islam, al-Kindi wrote in Baghdad on such subjects as meteorology, specific weights, geometrical optics, and physiology. Gaining access to Arabic translations of Aristotle's *Metaphysics* and having translated the *Ennaeds* of Plotinus and some of the works of Proclus, both of whom were Neoplatonists, he wrote a well-known treatise, *Theory of Aristotle*, interpreting his philosophy from a Neoplatonic point of view.

Al-Khwarizmi, who was born in Persia and studied Hindu mathematics but wrote in Arabic, composed two extremely influential Arabic works in mathematics, one entitled (in translation) *Arithmetic* and the other *Algebra*. In the first he introduced the Hindu numerical system with its place notation of numbers in the tens, hundreds, thousands, etc. (for example, 2011), since referred to as Arabic numerals, that greatly simplified calculations over Roman numerals. His second treatise, *Algebra* (*aljabr*), introduced algebraic functions so crucial for expressing scientific laws as equations, along with the term 'algorithm.'

Among other important scholars of the second half of the ninth and early tenth centuries were al-Battani and al-Razi, or Rhazes as the latter was known in the West. Working mainly in Baghdad studying the Arabic version of Ptolemy's astronomical treatise which acquired the Arabic title *Almagest*, al-Battani, the greatest of all Arabic astronomers, published tables of precise observations of the motions of the sun and the moon, along with more accurate measurements of the obliquity and precession of the equinoxes, using trigonometric functions that the Arabs developed from the Greeks. Abu Bakr al-Razi, according to Fakhry, was the "first system-builder in Islam" and the "only great Platonist of Islam," as well as the "greatest medical author and practitioner of the…ninth and tenth centuries…exhibit[ing] a profound veneration for Plato…as well as to the great Alexandrian doctor and philosopher Galen…." Singer adds that his greatest medical work, known as his "Comprehensive Book," "gathers into one huge corpus the whole of Greek, Syriac, and early Arabic medical knowledge, including descriptions of smallpox and measles…. His account of them is a medical classic." (op. cit., p. 148)

The first half of the tenth century includes al-Farabi, one of the most prominent philosophers who worked in Aleppo and Damascus. Again according to Fakhry, "[h]e was the first outstanding logician of Islam, who commented on or paraphrased the six books of Aristotle's *Organon*, together with the *Rhetoric* and the *Poetics*… and also wrote several original treatises on the analysis of logical terms, which had no parallels until modern times." In addition to his previous works on Aristotle, his study of his medical investigations led him to interpret his material cause as the imperceptible components of a medication which produced the formal cause as the curative power. He also contributed to Islamic Neoplatonism describing the "emanationist world-view of Plotinus." Active in the second half of the tenth century, Abu l-Wafa was known as one of the greatest mathematicians following al-Khwarizmi who wrote commentaries on the latter's mathematics as well as on Euclid and Diophantus.

The later tenth and early eleventh centuries are known for their technological advances. Ibn al-Hatham, or Alhazen as he is known in the West, in *The Treasury of Optics* rejected Ptolemy's notion of vision as caused by the eyes emitting visual rays to external objects which then transmits the object's image to the eye, for Aristotle's more plausible explanation according to which the sensory form of the object is projected to the eye and then transmuted by the lens into the perceptual image. According to Singer, "he discusses the propagation of light and colours, optical illusions and reflection, with experiments for testing the angles of reflection and of incidence" (p. 152), analogous to Hero of Alexandria. He also attempted to explain the nature of light, the rainbow, and how light is reflected from spherical and parabolic mirrors. Again as Singer asserts:

> His fundamental study *On the Burning-sphere* represents real scientific advance, and exhibits a profound and accurate conception of the nature of focusing, magnifying, and inversion of the image, and of the formation of rings and colours by experiments. The work is far beyond anything of its kind produced by the Greeks. (p. 153)

Another renown figure of the tenth and eleventh centuries was the Persian scholar Ibn Sina, or as he is called in the West, Avicenna. Particularly known for his contributions to philosophy and for his investigations in physics and astronomy, his *Canon of Medicine* also proved to be one of the most influential medical treatises of the Middle Ages owing to its synthesis of the biological writings of Aristotle, the anatomical and physiological investigations of Galen, and the medical research of the Arabs. His treatise *Healing* continued the Neoplatonic interpretation of Aristotle describing his planetary spheres and successive intelligences as emanations of the Prime Mover.

The last Muslim scholar to be discussed, Averroës as he is known in the West but whose Spanish name was Ibn Rushd, was not from Eastern Islam but was born of Spanish parents in Cordoba in the twelfth century. Following their conquest of Spain, the Muslims established a Library and Academy in Cordoba, along with the magnificent mosque, in the second half of the tenth century, enabling Cordoba to contend with Baghdad as the greatest center of Islamic research. Because of its proximity to Europe, it was positioned to enhance the passing of the treasure trove of Greek learning, along with the Arabic commentaries, to the newly established universities in Europe, mainly in the 12th and 13th centuries.

Acclaimed as one of the greatest Muslim philosophers, Averroës was called "the commentator," as Aristotle was known as "the philosopher," because of his Neoplatonic commentaries on Aristotle's philosophy. Rejecting the Genesis account of creation incorporated into the Koran that claimed the world had been created from nothing by God at a definite time, Averroës asserted that the universe was constantly being recreated but since the Prime Mover was eternal, so was his continuous recreation. He also defended the Neoplatonic conception that the human soul was of the same nature as the Divine soul. Though his books were burned by royal decree because of their unorthodox doctrines, they had a tremendous influence on later scholastic thought.

According to Fakhry, "it was thanks to the Latin translations of Ibn Rushd's [Averroës'] commentaries that the rediscovery of Aristotle in Western Europe and the concomitant emergence of Latin Scholasticism, one of the glories of late medieval thought, were made possible." While this may be true, credit must also be given to Thomas Aquinas' synthesis of Aristotelianism with Christianity in his great *Summa Theologica* in the 13th century that offered to the scholastics of the 14th and 15th centuries a more authentic interpretation of Aristotle's philosophy.

Regrettably, this remarkable revival of scientific inquiry began to wane from the 11th century due to the emergence of a more conservative and repressive Islam. The major reasons usually attributed to the decline of Arabic scientific and mathematical research are the following: first was the conquest of much of the Middle East by the Seljuk Turks in the middle of the eleventh century who had recently been converted to Islam. Extremely conservative in their religious views, they were opposed to scientific inquiry which led to their closing the Baghdad Academy. Second was the capture in the 12th century of Cordoba by the Spanish King of Castile, Ferdinand III, closing the greatest center of Muslim research in the West.

Third was the repressive religious influence of al-Ghazali, also in the 12th century, who rejected Muhammad's assertion that scientific inquiry would enhance religious beliefs, claiming instead that sacred beliefs were based on the revelations of the Koran and confirmed by mystical experiences, prayer, and faith, thus superior to scientific knowledge. Revered for his inspired preaching of Sufism, Sufic orders sprang up throughout Islam in the following century exerting a tremendous influence on Islamic culture—an influence renewed today by the resurgence of al Qaeda and the Taliban that are motivated by the illusory vision of recreating the Arabic Caliphate in the West.

The fourth reason was the barbaric invasion of the Mongols in the 13th century which devastated much of Islamic culture. Yet despite its limited history and somewhat early demise, civilization owes a great debt to the diverse Arabic speaking scholars for their preservation, enhancement, and transmission of Greek scientific and mathematical treatises. For by the 12th century translations into Latin of the original Greek and Arabic scientific and philosophical texts were widely disseminated which contributed to the Renaissance in the West. Unfortunately for the Muslims, their turn to conservative Sufism with its absolutistic beliefs excluded them from the liberating effects of science in promoting free inquiry and civil rights, accounting for their present backwardness in contrast to Western culture and hostility among the *jihadi*.

Prior to the Arabic revival of Greek science, the Greek philosopher prominent in the early development of Christianity was Plato. This was because he disparaged empirical beliefs derived from the imperfect contents of the Receptacle, instead extolling the abstract methodology of mathematics since its pursuit freed the intellect from its dependence on the senses to attain knowledge of the Ideal Archetypical Forms culminating in the Form of the Good, a philosophy more congenial to the Christian transcendental perspective. In contrast, as just recounted the second millennium was inspired by the investigations of the Arabs which had been stimulated by translations of the ancient Greek manuscripts, particularly Aristotle but also Hippocrates, Galen, Hipparchus, Ptolemy, and Plotinus.

CHAPTER THREE

The Early Scholastics and Their Revisions of Aristotle's Philosophy

Despite the demise of Arabic science, the transmission of Greek and Arabic manuscripts to the West after being translated into Latin in the 12th and 13th centuries, coinciding with the establishment of universities in Paris, Bologna, Oxford, Cambridge, Padua, and Prague, stimulated the resumed interest in scientific and mathematical inquiries in those institutions. The consequent reorientation in the West is exemplified in an imaginary conversation with his nephew related by the 12th century English scholastic philosopher, Adelard of Bath, in his *Questiones naturales*. As described by Crombie, during their imagined discussion his nephew declared (as Augustine had previously) that a certain natural event could be explained only as "a wonderful effect of the wonderful Divine Will," to which Adelard replied that "while it was certainly the Creator's will that it should happen...it is also not without a natural reason, and that this was open to human investigation."[18] As commonplace as this may sound, it marks a major turning point in that it implied that simply attributing something to God's will was no longer considered a sufficient explanation.

But acknowledging the significance of empirical investigations was not so simple considering the conceptual adjustments this required: supplementing (or replacing) God's Will by natural causes, theology by metaphysics, faith by reason, miracles by natural explanations, revelations by rational demonstrations, dogma by tested hypotheses, eternal truths by probabilistic explanations, and church authority by scientific inquiry. As Crombie states: "The history of science shows that the most striking changes are nearly always brought about by new conceptions of scientific procedure. The task demanding...the revision of the questions asked, the types of explanation looked for, the criteria for accepting one explanation and not another." (p. 1) How different this is from accepting as eternally true such implausible beliefs as the virgin birth of Jesus, the transubstantiation of the bread and wine at the Eucharist into the flesh and blood of Jesus while retaining their ordinary properties, and the Trinitarian doctrine that the Godhead consists of three consubstantial persons, the Father, the Son and the Holy Spirit, the latter two doctrines not even attributed to divine revelation but to decisions taken at several Church Councils.

The scholastics not only confronted a diversity of beliefs, but also the task of accepting a more realistic understanding of the universe and selecting which among the philosophical systems, Plato's, Aristotle's, Plotinus', Avicenna's, Averroës', or Aquinas,' should be chosen as the basis of the new worldview. And since they later generally considered the treatises of Aristotle as authoritative, this meant having to assimilate his more empirical philosophy for which they had little preparation, most of their training having been in the standard curriculum of the *Trivium* and Quadrivium of the emerging medieval universities, rather than in the specific scientific disciplines, except for astronomy.

Moreover, in contrast to religious dogma which is considered infallible and eternally true, once a naturalistic framework for interpreting nature is attained further investigations nearly always requires modifying the framework to accommodate the new evidence, the task faced by the founders of modern classical science. The irony is that though Aristotle's philosophy made this critical examination possible because it accommodated the predominant *observable* features of and causes in the world, eventually it had to be *entirely* rejected for a radically different method of inquiry and explanatory system. This new approach incorporated exact observations, unexpected data derived from new technologies such as the telescope, microscope, and spectroscope, experimental testing of hypotheses, and newly discovered laws. Moreover, there came the realization that "the language of nature was mathematics," in that the properties of the microworld, in contrast to the macroworld, were depicted in magnitudes such as mass, energy, force, charge, and velocity, not sensory qualities. To understand this contrasting transition some understanding of Aristotle's mode of explanation and cosmological system is required.

Although nature presents a continuous process of change and becoming, it also displays stability and order that Aristotle explained in terms of principles derived from observable inductions accounting for "why" they occurred, as he states in the *Physics*.

> Knowledge is the object of our inquiry, and men do not think they know a thing till they have grasped the "why" of it.... So clearly we must do this as regards both coming to be and passing away and every kind of physical change, in order that, knowing their principles, we may try to refer to these principles each of our problems.[19]

He then describes the underlying or inherent causes as four in number : (1) the material cause: "that out of which a thing comes to be and which persists, is called 'cause,' e.g., the bronze of the statue, the silver of the bowl, and the genera of which the bronze and the silver are species." (p. 240). This cause is relative since the bronze itself is composed of the alloys tin and copper which are finally resolved into an irreducible "prime matter." (2) The formal cause: "the form or the archetype, i. e., the statement of the essence, and its genera, are called 'causes'...." (p. 240) Here again he refers to the "genera" which will be crucial to his mode of explanation. (3) The efficient cause: "the primary source of the change or coming to rest...and generally what makes of what is made and what causes change of what is changed" (p. 241), the one closest to modern science. (4) And lastly the final cause (derived from his biological investigations) "in the sense of end or 'that for the sake of which' a thing is done...." (p. 241) This final cause is intrinsic to the essence or genus that determines what things change into and the reason Aristotle's system is considered teleological and described as "organismic."

Although both Aristotle and modern scientists would claim that an object's causal powers depend upon its essential nature, they differ totally in their conceptions of what that nature is. While Aristotle's explanatory schema mainly reflects the influence of Plato's Forms and his investigations of biological phenomena, the theoretical framework of modern scientists is based on the discovery of microphysical or submicroscopic entities. For Aristotle it is the "form," "archetype," "species," "genus," or "definition" of the object that constitutes its inherent nature. Thus his crucial question is: "How then by definition shall we *prove* substance or essential nature?" (p. 165)

In contrast, the causal powers of modern science are dependent on an object's atomic-molecular structure, charges, and forces. As the conceptual framework of modern science evolved each of Aristotle's causes had to be transformed: his material cause as an underlying "substance" or "prime matter" reinterpreted as subatomic particles such as electrons, protons, and neutrons with electrical properties and forces; the formal as "form" or "genera" redefined as their atomic number or weights producing their macroscopic properties; the efficient not in terms of its genus or kind but the various forces that produce the changes; and the final which is eliminated except in the case of biological organisms.

Unlike Plato who *denied* that true knowledge could be derived from sensory observations but depended on detaching oneself from the senses and with the aid of mathematics intuit the independently exisiting Forms that are the ideal archetypes of empirical objects, Aristotle believed that though mathematics could be used to describe abstract magnitudes and geometrical shapes, it could not be applied to the concrete sensory world with the exception of optics. Replacing mathematics in *The Prior Analytics* with the formalism of sentential deductive logic that he had created, his method of scientific explanation consisted of deducing the nature or existence of an essential property of an entity or occurrence from true universal premises or genera arrived at inductively that included that property.

As his classic examples illustrate: if one asks "why Socrates is mortal?", the answer consists in demonstrating that the property mortality is essential to Socrates by the syllogism that "All men are mortal, Socrates is a man, therefore Socrates is mortal;" or showing "why the planets do not twinkle" (in contrast to the stars) by the deductive argument that "No proximate celestial body twinkles, the planets are proximate celestial bodies, therefore the planets do not twinkle;" or answering "Why the moon suffers eclipse?" by the inference that "The privation of the moon's light by the interposition of the earth causes the eclipse." (p. 160)

By considering the middle terms "man" and "twinkles" connecting the premises in the above deductions as the *causal* determinations of the conclusions, he was able to view these *verbal* definitions and logical deductions as *causal* explanations and empirical demonstrations. This is evident in his explication of the moon's eclipse, even though it is not stated as a syllogistic demonstration. As he concludes: "in all our inquiries we are asking either whether there is a 'middle' or what the 'middle' is: for the 'middle' here is precisely the cause, and it is the cause that we seek in our inquires." (p. 159) Thus his explanations differ from those of later science because they depend upon inferring the genera or species of things from their *observable* features, while those of modern science consist of discovered laws usually explained by the entities atomic-molecular structures and microphysical properties.

His actual investigations, however, proved more effective and lasting than this classificatory methodology, as indicated in Darwin's effusive praise of his biological investigations and as can be seen in his cosmological system that was still defended by the scholastics at the time of Newton. The latter is divided into two realms, the celestial and the terrestrial, as described in the *De Caelo* (*On the Heavens*), *Physics,* and *Metaphysics.* The celestial realm consisted of the seven planets, Moon, Mercury, Venus, Sun, Mars, Jupiter, and Saturn, in successive orbits ascending from the central stationary earth. They are carried in their individual circular orbits rotating around the earth with two uniform motions. Each planet shares in the diurnal rotary motion from east to west of the eighth sphere of the fixed stars constituting the *daily rotation of the entire universe*. In its second motion each planet again revolves uniformly, but with its own speed, from west to east to complete its *particular orbital revolution*. Though modified by Ptolemy with his epicycles, the assumption of *uniform circular motions around a central earth* was not rejected until Copernicus proposed his heliocentric system and Johannes Kepler introduced *nonuniform motions in elliptical orbits circling the sun* to "save the phenomena," in Plato's phraseology.

This celestial realm, though composed of a weightless, incorruptible, eternal aither whose inherent uniform circular motion would seem to be self-sustaining, yet required "Secondary Movers" or "Intelligences" to direct each of the planets in their different orbital velocities. The number of planetary spheres was increased to forty-seven or fifty-five (the exact number is contentious) to include counteracting spheres to negate the transmission of their motions to their adjacent spheres.

While this explanation would seem to be sufficient, in the *Metaphysics* Aristotle argues that the celestial spheres owe their motions ultimately (though not temporally) to the *Primum Mobile* (Prime Mover) that exists beyond the cosmos in the non-spatial outermost empyrean realm. To avoid an infinite regress due to the premise that whatever moves presupposes a mover, he introduced the Prime Mover as the ultimate source of the motion of the entire cosmos, not by contact, but by an empathetic attraction toward it. As he states in the *Metaphysics*:

> …since that which is moved and moves is intermediate, there is something which moves without being moved, being eternal, substance, and actuality. And the object of desire and the object of thought move in this way; they move without being moved…. The final cause [the Prime Mover], then, produces motion as being loved, but all other things move by being moved. (p. 879; brackets added))

This explanation apparently is based on the primitive animistic assumption that only a psychic agency or intelligent being can be an ultimate cause of movement.

In contrast to the unchanging perfect nature of the celestial world, the terrestrial consists of the four Empedoclean elements, fire, air, earth, and water, that undergo continuous interaction, change, destruction, and reconstitution caused by natural forces such as fire, wind, and waves. Each element has an inherent rectilinear motion either upward or downward, fire and air naturally rising and water and earth naturally falling. If not disturbed each would be at rest in its natural place, fire just below the innermost celestial sphere, earth in the center of the cosmos, water on the surface of the earth, and air in between. As this is not the case, he adds that "[w]hen these things are in motion to positions the reverse of those they would properly occupy, their motion is violent…[or] unnatural…." (p. 364; brackets added)

He explained the fall of material objects after being projected upward as due to their natural tendencies to reach the earth, with their accelerations variously explained as dependent upon their inherent weights, the pressure of the air above them, their distances from the earth, or the resistance of the air below. Because horizontal motion is not natural to objects, projectile motion posed a special problem which he explained as due to the action of the mover compressing the air behind the object which, when expanding, propels the object but is also opposed by its resistance before it.

The inadequacies of these explanations of motion, whether the uniform circular motion of the celestial bodies, the accelerated upward and downward motions of terrestrial elements, and the unnatural or violent motion of displaced elements or projectiles, would constitute the major theoretical investigations of the scholastics of the late Middle Ages, to which we turn next, along with the founders of modern classical science. But as indicated previously, even though mistaken Aristotle's significance consists in his having introduced these physical or natural explanations in a theoretical framework which was subject to critical analysis, tests, and revision, the essential features of modern scientific inquiry, though not of his. This is why his explanatory system, not Plato's, was mainly adopted by the later scholastics and considered the final truth until refuted by the modern scientific revolution of the 16th and 17th centuries.

CHAPTER FOUR

Contributions of Grosseteste and Later Scholastics

Although Marshall Clagett's *The Science of Mechanics in the Middle Ages* contains the most comprehensive collection of the original writings of the later Scholastics, I shall rely on the more accessible description by Crombie to show how the scholastics initiated revisions in Aristotle's cosmology that prepared the way for, but did not actually attain, the required solutions achieved by the founders of modern classical science beginning with Copernicus. According to Crombie (in the book cited previously) in was Robert Grosseteste, Bishop of Lincoln who, in the 13th century founded a distinguished school of scientific thought at Oxford University that led the renewal of scientific inquiry owing to the availability of the following works of Aristotle and others.

> By the end of the twelfth century not only the *Analytics*, but also the *Physics, De Generatione et Corruptione, De anima Parva Nautralia,* and the first four books of the *Metaphysics* of Aristotle had been translated into Latin from the Greek, and the *De Caelo* and the first three books of the *Meteorologica* had been translated from the Arabic. During the same period Latin versions from either Greek or Arabic, and sometimes from both, were made of the *Elements, Optics,* and *Catoprics* of Euclid ... part of the *Conics* of Apollonius, the *Almagest* and *Optics* of Ptolemy...and numerous works by Hippocrates and Galen. (pp. 35–36)

But it was Aristotle's works especially that provided the background for Grosseteste's investigations and those of his scholastic followers at Oxford, such as Roger Bacon, John Pecham, Duns Scotus, William of Ockham, Thomas Bradwardine, and and John of Dumbleton. Following Grosseteste in his teachings of Aristotle Albert Magnus, known as Saint Albert the Great, when teaching at the University of Paris and various German institutions, was especially influential in spreading the latter's philosophy in Europe. Thomas Aquinas, an Italian philosopher and theologian who was the favorite student of Albert Magnus at Paris and perhaps the greatest of the 13th century scholastics, taught theology at Paris and later went to Naples where he established an Institute of Studies. He is best known for his thesis that faith and reason are harmonious making possible a synthesis of Aristotelianism with Christianity.

Yet despite these other centers of Aristotelian studies, Crombie asserts that it was mainly from the time of Grosseteste that

> the experimental science which has ever since been an essential part of the Western world began to appear in centre after centre. It is found from the end of the thirteenth century in the Germanies and in the medical schools of Padua and other north Italian towns. There it was taken up in the fifteenth and sixteen centuries by Leonardo de Vince and the Italian physiologists and mathematical physicists. And so it went to Galileo, William Gilbert, Francis Bacon, William Harvey, Descartes, Robert Hooke, Newton, Leibniz, and the world of the 17th century. (pp. 14–15)

Grosseteste's own contribution is evident in his modification of Aristotle's methodology to make it conform more to the emerging practice of scientific inquiry. Thus he replaced Aristotle's genera and species, along with his definitions as the premises of scientific explanations, by laws describing empirical correlations. Renaming Aristotle's "induction and deduction" as "resolution and composition," he relied on *experimental tests* to determine which laws should be selected as the correct premises for explaining phenomena. He introduced the principle of *reductio ad impossible*, or as it is known today, *reductio ad absurdum*, as a method for refuting erroneous hypotheses by showing that they led to false conclusions. (Cf. p. 84) He also rejected Aristotle's claim that apart from optics mathematics was of no use in science because it applied only to abstract configurations or quantities, not actual physical properties, maintaining instead that mathematical quantities existed as quantitative aspects of physical things: "quantitative dispositions are common to all mathematical sciences …and to natural science" (p. 91), and therefore is indispensable to scientific inquiry.

Again contrary to Aristotle Grosseteste realized that discovering the underlying causes producing the correlations in nature also was necessary, although he added that inferring from an effect its particular cause was more uncertain than ascertaining the effect of a cause because there may be several causes of the effect. (Cf. p. 81) Perhaps influenced by Augustine, he rejected Aristotle's (and/or Plato's) recourse to metaphysical knowledge to achieve certainty claiming, in the words of Crombie, "that there was no certainty in metaphysics since it is by Divine illumination and by this alone man could have certain knowledge of the essence of the real." (p. 131) Moreover, "'[s]ince...the truth of each thing is the conformity of it to its reason in the eternal Word, it is evident that every created truth is seen only in the light of the supreme truth.'" (p. 131) Needless to say, this latter belief was not a conception congenial to the development of science. As I wrote elsewhere, "when science is brought in, God is eventually left out."

Apparently suggested by his optical investigations, this doctrine of Divine Illumination was reinforced by his Neoplatonic conception of the origin of the universe known as the "metaphysics of light." If one excludes the role of God and equated this light with radiation, with some charity it is possible to see in his view a hint of the modern theory of the creation of the universe as originating from a singular burst of energy, known as the Big Bang, whose explosion spewed forth the expanding radiation which, as it cooled, created the three-dimensional universe with all its physical diversity. As Crombie describes Grosseteste's account, God first created (Aristotle's) unformed prime matter (*materia prima*), then light (*lux*) which, by diffusion, originated its three dimensions transforming it into corporeality, along with endowing it with motion. In Grosseteste's own words:

> The first corporeal form...I hold to be light. For light (*lux*) of its own nature diffuses itself in all directions, so that from a point of light a sphere of light of any size may be instantaneously generated.... Corporeality is what necessarily follows the extension of matter in three dimensions, since each of these, that is corporeality and matter, is a substance simple in itself and lacking all dimensions.... And, in fact, it is light, I suggest, of which this operation is part of its nature, namely, to multiply itself and instantaneously diffuse itself in every direction. (p. 106; brackets added)

Grosseteste of course was mistaken in thinking the universe as it now exists had been created instantaneously since we now have evidence that the universe is about 13.7 billion years old. He also attempted to explain more specific phenomena, such as Aristotle's celestial orbits and the tides, along with optical phenomena like reflection, refraction, colors, and rainbows. The range is truly extensive, justifying Crombie's praise of Grosseteste as a pivotal influence in the 13th century renewal of scientific inquiry which brings us to the contributions of his followers at Oxford and Paris after his death in 1253.

It was Roger Bacon (ca 1219-1292) "who most thoroughly grasped, and who most elaborately developed Grosseteste's attitude to nature and theory of science." (p. 139) Like Grosseteste he anticipated Galileo in also emphasizing the importance of mathematics in scientific explanations, declaring that it was the "door and key of the sciences and things of this world...[thus] we must place the foundations of knowledge in mathematics." (p. 143; brackets added). He corrected Grosseteste's notion that light is propagated instantaneously, even though he admitted we do not have perceptible evidence of its finitude. His research in vision was especially acute, Crombie asserting that "his account of vision was one of the most important written during the Middle Ages and it became a point of departure for seventeenth-century work." (p. 151)

Other important scholars of the 13th century were John Pecham and Witelo whose research in optics was especially influential, along with John Duns Scotus who Crombie states "absolutely rejected Grosseteste's position that man could know nothing about the world without Divine illumination, a view he held led to scepticism...." (p.168) Instead he invoked the principle of the uniformity of nature declaring that "whatever occurs in a great many cases from some cause...is the natural effect of that cause" (p. 170), so that in some cases a sufficient number of confirmations would justify believing in it with certainty, a precept that goes beyond modern science.

The influence of Grosseteste and Roger Bacon was not restricted to Oxford, but also extended to astronomical research in France in the 13th century. At the time there was considerable controversy between the Aristotelians and the Ptolemaists as to which astronomical system could best "save the appearances," the Franciscan astronomer Bernard de Verdun strongly claiming that the astronomical evidence supported Ptolemy. Then about 1310 Pietro d'Abana, in his *Lucidator Astronomiae*, as Crombie states, "introduced the radical position that the heavenly bodies were not borne on spheres but moved freely in space, followed by the even more enlightened view that the appearances could be saved most simply by considering the earth instead of the heavens to be in motion." (p. 202) Furthermore, Crombie asserts that this initial dismantling of Aristotle's cosmological system was reinforced in the late 14th century by the remarkable Nicole Oresme "who also proposed discarding the celestial orbs and accept that the planets had a relative motion in an infinite, geometrical space." (p. 202)

Other aspects of Aristotle's explanatory system also were being critically replaced in the 14th century. Rejecting Aristotle's claim that all projectile motion had to be caused by direct contact with the mover, William of Ockham cited sunlight and magnetic attraction as evidence of action at a distance, a conception supported later by Gilbert's magnetic and electrical experiments. Ockham also dismissed Aristotle's final cause arguing that since it would have to produce its effect retroactively "when it does not exist...this movement towards an end is not real but metaphorical." (p. 174) This was an important advance because it tended to replace intentional explanations by physical causes.

Another major contributor to the advance of science in the 14th century was Thomas Bradwardine who Crombie asserts was the real founder of the school of scientific thought associated with Merton College, playing the same role at Oxford in the 14th century as Grosseteste did in the 13th century. (p. 178) In his *Tractatus Proportionum* investigating what is called Aristotle's "Peripatetic law of motion " that the velocity of a propelled object is due to the magnitude of the mover less the resistance of the air, he attempted to express mathematically the functional relationship among the magnitudes involved: how *change* in velocity was related to power and resistance. (Cf. p. 179)

In his formulation he was innovative in having letters of the alphabet represent different quantities, but described the operations of addition, subtraction, and division verbally, rather than with symbols such as +, −, ÷. Nonetheless, as Crombie states,

> his formulation of the problem in terms of an equation, and one in which the complexity of the relationships involved was fully recognized, was an original and important contribution to mathematical physics in general and to dynamics in particular. Through him fourteenth-century natural philosophers, both in Oxford and Paris, got the beginnings of a conception of the the use of mathematical functions in physics and fifteenth-century Italy saw the beginning of an experimental investigation not only of his dynamical function but also of the Peripatetic law of motion itself. (p. 180)

Not only did Aristotle's system not include functional relations, it did not contain clear conceptions of average motion, acceleration and deceleration, or instaneous motion which were prerequisites for any precise description of the various natural motions. However, according to Maurice Clavelin, "by 1320 Oxford physicists were treating velocity as an intensive magnitude, subject to intension and remission…and they went on to distinguish the quality of a motion (that is, its velocity) from its quantity (that is, the distance traversed)."[20]

The distinction between uniform and nonuniform motions was probably introduced by John of Dumbleton who, in his *Summa Logice et Philosophie Naturalis*, distinguished between latitudes of change (*intensio*) from the distance or time (*extensio*) of motion. Owing to the writings of Dumbleton, William Hytesbury, and Richard Swineshead (called the Calculator) a consensus regarding the distinctions and descriptions of these various motions began to emerge. Despite the unusual, unfamiliar terminology the conceptual clarifications are evident in Crombie's summary.

> A change was said to be "uniform" when, in uniform local motion, equal distances were covered in equal intervals of time, and "difform" when, in accelerated or retarded motion, increments of distance were added or subtracted in successive intervals of time. Such "difform" change was said to be "uniformly difform" when equal increments were

added or subtracted in successive intervals of time, and "difformly difform" when unequal increments were added or subtracted. (pp. 181–182)

A further clarification was introduced by Oresme who injected the term *"velocitatio"* to designate a continuous increase of velocity which at least approximated our meaning of acceleration. As Clavelin claims, "while it did not yet describe *the rate of change in velocity*, in the sense that classical mechanics later defined it, Oresme's *velocitatio* nevertheless referred to variations in speed as such, and hence transformed the latter into a distinct object of thought." (p. 69). Moreover, Oresme's conception of speed as an intensive magnitude that changes from moment to moment implied that the velocity at any one moment had a definite or instantaneous intensity which, according to Clavelin, "must therefore be counted among the major contributions of medieval science." (p. 70) Yet the essential expression of instantaneous velocity as the differential coefficient of space with respect to time (*ds/dt*) as their values approach zero awaited the development of calculus by Newton and Leibniz in the 17th century. As important as these mathematical advances were, they still lacked the final formulations or refinements provided by that great century of scientific achievements.

Nonetheless, it has been argued that at least the "theory of impetus," introduced by Jean Buridan in the 14th century to solve one of the weakest explanations of Aristotle, projectile motion, was a precursor of the later theory of inertia. As described previously, Aristotle contrasted natural motions, whose movement was inherent, to the unnatural or violent motion of projectiles that required an external cause *in constant contact* to explain their *continued motion* when the original mover ceased—in the example of the projectile the decompressing air behind it. The difficulty was how the force of the air on such a small surface as the end of a spear or a stone could propel it while the air in front retarded it.

Buridan's solution was to dispense with the assumption that the continued motion required the *constant contact* of a mover, proposing instead that the action of the mover *impressed an impetus in* the object that continued the propulsion after the contact ceased, the resistance of the air and tendency to fall to the earth eventually overcoming the impetus. As proposed by Buridan and quoted by Clavelin:

> "Therefore, it seems to me…that the motor in moving a moving body impresses in it a certain *impetus* or a certain moving force…in the direction toward which the mover was moving…. And by the amount the mover moves that body more swiftly, by the same amount it will impress in it a stronger impetus. It is by that impetus that the stone is moved after the projector ceases to move. But that impetus is continually decreased by the resisting air and by the gravity of the stone, which inclines it in a direction contrary to that in which the impetus was naturally predisposed to move it." (p. 92)

Yet it is questionable whether the concept of impetus was exactly equivalent to that of inertia because according to Newton's first law of motion "*Every body continues in its state of rest, or of uniform motion in a right line, unless it is compelled to change that state by force impressed upon it.*"[21] For Buridan it seems that the impetus must *continue* to act on the object to keep it in motion while for Newton once the object is set in motion its *inertial* movement continues naturally, without the necessity of an added force, until overcome by the impeding effects of air and gravity. Rather than *sustaining* its continued motion, any added force would *increase* the velocity of the object.

This is an excellent example of why scientifically overcoming past explanations is often so difficult because the newly discovered explanatory causes initially seem counterintuitive or to be contrary to ordinary experience. Given the usual tendency of objects to come to rest after being propelled, it does not appear reasonable to assume that they would naturally continue in motion except for the opposition of external forces, as the law of inertia states. Thus as close to the concept of inertia as Buridan was, and though he replaced Aristotle's air compression with an impetus, like Aristotle he apparently assumed that the continued effect of impetus was necessary to explain the continuous motion of an object after the original cause has ceased.

Still, as we have seen, the scholastics can be credited with the initial revision of Aristotle's cosmological and explanatory system replacing it preliminarily with a more advanced scientific framework including the following methodological and conceptual changes: the replacement of Aristotle's deductive method of explanation in terms of genera and species with the discovery of mathematical laws based on quantitative properties incorporating functional correlations; the use of experimentation to supplement observations and to test hypotheses; the rejection eventually of divine illumination as the ultimate source of knowledge; the awareness that miracles were inconsistent with the uniformity of nature, justifying the search for purely physical laws; the percipient explanation of the cause of the physical universe in terms of light radiation; the realization that the earth moved and that the celestial bodies or planets were not borne on spheres but moved freely in space; and that magnetism and electricity were evidence of forces acting at a distance that challenged Aristotle's assumption that any unnatural motion required direct physical contact; the introduction of the concept of impetus anticipating the theory of inertia; the rejection of intentional final causes for purely physical explanations; and the introduction of more precise mathematical terms to differentiate uniform from nonuniform motion, velocity as an intensive magnitude apart from the distance covered, and implicit concepts to identify acceleration as well as instantanous motion. Though partial, these developments were signs of progress.

All of these conceptual advances are justification for Thomas Kuhn's incisive assessment of the achievements of the scholastics.

> The centuries of scholasticism are the centuries in which the tradition of ancient science and philosophy was simultaneously reconstituted, assimililated, and tested for adequacy. As weak spots were discovered, they immediately became foci for the first effective research in the modern world. The great new scientific theories of the sixteenth and the seventeenth centuries all originate from rents torn by scholastic criticism in the fabric of Aristotelian thought. Most of those theories also embody key concepts crecreated by scholastic science.[22]

CHAPTER FIVE

Copernicus, Kepler, and Gilbert

Despite the progress of the scholastics in revising Aristotle's cosmological framework, Nicolas Copernicus's publication of *De revolutionibus orbium coelestium* in 1543 is generally regarded as initiating the revolutionary changes essential for creating modern classical science. Although heliocentrism had been introduced by Aristarchus in the third century BC and was supported by Pietro d'Abana and Nicole Oresme in the 14th century, it was not adopted as a seriously contending hypothesis until after Copernicus's publications. Even so it has been argued that just replacing the earth by the sun at the center of the cosmos and attributing two motions to the former, an annual revolution around the sun and a diurnal rotation on its axis, while retaining the rest of Aristotle's system of concentric spheres along with their uniform motions was not that revolutionary. But there are three reasons why I believe this is not true.

The first is that rejecting the homocentric universe was still considered heretical. This is reflected in the harsh response of Martin Luther, a contemporary, to Copernicus's *Commentariolus* (*Little Commentary*), his preliminary version of the heliocentric theory written in 1630.

> "People give ear to an upstart astrologer [Copernicus] who strove to show that the earth revolves, not the heaven or the firmament.... This fool wishes to revise the

entire science of astronomy; but sacred Scripture tells us that Joshua [Joshua10:13] commanded the sun to stand still...not the earth."[23] (Brackets added)

This again illustrates how wrong the defenders of the Bible have been.

The second is that though several of the scholastics were willing to admit that the greater simplicity and harmony of the heliocentric system outweighed the strong evidence of ordinary experience, yet owing to the creation account in Genesis and Aristotle's support geocentrism was still defended by the majority of those who counted. The third is that the innovations of the scholastics mentioned did not coalesce into a convincing alternative theoretical framework. The fourth is the fact that Kepler would not have been motivated to discover his three laws of planetary motion and introduction of their causal forces had he not been attracted by Copernicus's sun centered theory, a fact that offered other possible explanations. So if the innovation itself was not especially revolutionary, it was "revolutionary making" as claimed.

One could hardly find a more unlikely person to initiate a revolutionary saga than Copernicus (1473-1543). Extremely timid by nature, it is largely owing to the persistence of Rheticus, an Austrian who was a young professor of mathematics and astronomy at the University of Wittenberg (also Luther and Hamlet's university), that his *magnus opus*, the *De revolutionibus orbium coelestium* (*On the Revolutions of Heavenly Spheres*), was brought to publication. As mentioned previously, Copernicus had written a preliminary version of his heliocentric theory in 1530 entitled *Commentariolus*, but owing to his fear of ridicule, illustrated by the reaction of Luther, he did not publish it but circulated it among a few friends. Encouraged by their favorable reaction, he began work on the *De Revolutionibus*, as the title is usually abbreviated.

Learning of this Rheticus took leave of absence from the university in the spring of 1539 to persuade Copernicus to have it published. Rebuffed at first, about a year later he did convince him to have it published staying with him to help prepare the manuscript for printing, reading the entire work, and even recopying it by hand to present to the printer in Nüremberg, although when forced to leave Wittenberg (allegedly because of homosexual accusations), his colleague Osiander supervised the final printing. Copernicus was convinced that by replacing the earth with the sun at the center of the universe and attributing two motions to the former he could derive a simpler, more harmonious, astronomical system than Ptolemy's. In his Preface and Dedication to Pope Paul III in an attempt to gain his approval, he states that

> nothing except my knowledge that mathematicians have not agreed with one another in their researches moved me to think out a different scheme of drawing up the movements of the spheres of the world.... Moreover, they have not been able to discover or to infer the chief point of all, *i.e.*, the form of the world and the certain

commensurability of its parts. But they are in exactly the same fix as someone taking from different places, hands, feet, head, and other limbs—shaped very beautifully but not with reference to one body and without correspondence to one another—so that such parts made up a monster rather than a man.[24]

He continues by asserting that since others in the past had proposed the heliocentric view as a more adequate explanation of the planetary orbits for similar reasons, he felt justified in doing the same.

> Therefore I also…began to meditate upon the mobility of the Earth. And although the opinion seemed absurd, nevertheless because I knew that others before me had been granted the liberty of constructing whatever circles they pleased in order to demonstrate astral phenomena, I thought that I too would be readily permitted to test whether or not, by the laying down that the Earth had some movement, demonstrations less shaky than those of my predecessors could be found for the revolutions of the celestial spheres. (p. 6)

He then presents the favorable theoretical conclusions that can be obtained from making these assumptions.

> And so, having laid down the movements which I attribute to the Earth farther on in the work, I finally discovered by the help of long and numerous observations that if the movement of the other wandering stars [planets] are correlated with the circular movement of the Earth, and if the movements are computed in accordance with the revolution of each planet, not only do all their phenomena follow from that but also this correlation binds together so closely the order and magnitudes of all the planets and of their spheres or orbital circles and the heavens themselves that nothing can be shifted around in any part of them without disrupting the remaining parts and the universe as a whole. (p. 6; brackets added)

He adds that since "[m]athematics is written for mathematicians" and knowing of the Pope's "love…of mathematics," he would leave "what I have accomplished in this matter…to the judgment of Your Holiness in particular and to that of all other learned mathematicians." (p. 7) Appealing to the Pope not as an ecclesiastical authority, but as a mathematician could be seen as a turning point since it implied that any decision should be based on the astronomical evidence and not on scripture, yet this did not prevent the later disgraceful treatment of Galileo by Pope Urban VIII and the Inquisition.

Though his *De Revolutionibus* is filled with diagrams, tables, and astronomical arguments showing that when the heliocentric hypothesis is adopted, as he stated above, not only do all the astronomical phenomena fall in place, but the resultant system is so tightly integrated that nothing can be altered without disrupting the entire structure. Yet, apart from offering the prospect of reforming the Ecclesiastical Calendar, he was unable to formulate the exact astronomical laws depicting the "order and magnitudes of the planets and of their spheres or orbital circles" crucial for convincing the opposition. The latter was left to Kepler whose discovery of the three astronomical laws and physical forces governing planetary motion was a critical development, along with Galileo's telescopic discoveries!

Turning then to Kepler, it is dreadful that the life of such an exemplary person was marked by the tragic death of his first wife and four children, one from his first marriage and three from the second, constant financial struggles, and the preposterous charge of witchcraft brought against his aged mother who was only saved due to her unwavering assertion of her innocence, despite the threat of torture and death, and by Kepler's defense aided by influential friends at court. Yet these painful experiences were somewhat compensated for by his extraordinary intellectual gifts leading to his remarkable astronomical discoveries which helped demolish the geocentric system and create modern classical science.

Like other geniuses, Kepler recalled events in his childhood, such as seeing when he was six years old a great comet and at age nine a lunar eclipse, that influenced the direction his later life would take. But it was not until he was appointed professor of mathematics at the Protestant seminary at Graz at age twenty-three, when studying Copernicus's heliocentric system, that he developed a particular interest in astronomy. What attracted him was the realization that accepting the sun's position in the center of the universe offered the possibility of explaining the exact relative distances and sizes of the planetary orbits as due to the sun's influence.

His initial inquiry began by asking why there were just six planets, Mercury, Venus, the Earth (with the Moon as its satellite), Mars, Jupiter, and Saturn, with their unique orbits and distances from the sun and why the more distant planets moved more slowly and in what proportion to the distance? This is an excellent example of how revisions in a system open up unforeseen questions and possibilities of explanation. His initial attempt was guided by Plato's explanation in the *Timaeus* of how the four elements, fire, air, earth, and water, plus the universe itself, could be represented by the five Pythagorean geometric solids listed in the order of their creation: (1) fire: the tetrahedron (pyramid) formed by four equilateral triangles; (2) air: the octahedron formed by eight equilateral triangles; (3) water: the icosahedron formed by twenty equilateral triangles; (4) earth: the cube formed by six squares, and (5) the cosmos: the dodecahedron formed by twelve pentagons.[25] While Plato had introduced this scheme to account for the interaction of the four terrestrial elements or solids, Kepler thought he found in his account a possible explanation also of the structure of the cosmos itself.

A Christian Platonist believing that any geometrical order must preexist in God's mind, he was convinced that there being just six planets could not be accidental, but must have an intentional explanation, as stated in the *Mysterium Cosmographicum*.

> It is my intension…to show in this little book that the most great and good Creator, in the creation of this moving universe, and the arrangement of the heavens, looked… to those five regular solids, which have been so celebrated from the time of Pythagoras and Plato down to our own, and that he fitted to the nature of those solids, the number of the heavens, the proportions, and the law of their motions.[26]

A clear description of how these Pythagorean polyhedra and orbs are fitted around the five planets to form the structure of the cosmos, along with an exact diagram on the opposite page, is given by J. V. Field in his *Kepler's Geometric Cosmology*.

> It turns out that the dimensions of these planetary orbs are such that if a cube is inscribed in the inner surface of the orb of Saturn then its insphere will be the outer surface of the orb of Jupiter, and if a tetrahedron is inscribed in the inner surface of the orb of Jupiter then its insphere will be the outer surface of the orb of Mars, and if a dodecahedron is inscribed in the inner surface of the orb of Mars then its insphere will be the outer surface of the orb of the Earth, and if an icosahedron is inscribed in the inner surface of the orb of the Earth then its insphere will be the outer surface of the orb of Venus, and if an octahedron is inscribed in the inner surface of the orb of Venus its insphere will be the outer surface of the sphere of Mercury. Kepler presents this system in the way just described – working inwards from the sphere of Saturn so that we are alternately inscribing a regular polyhedron in a sphere and inscribing a sphere in a regular polyhedron.[27]

Although initially convinced by these polyhedral representations of the structure of the cosmos, Kepler was forced to acknowledge that his diagram did not conform to the astronomical evidence. Open minded enough to reject this model, he still retained his belief that the orbital dimensions were correlated with the centrality and influence of the sun, *so that whatever forces issued from the sun were the key to the explanation.* As we shall find, what is truly remarkable about Kepler, especially considering the strength of his Christian and Neoplatonic beliefs, was his ability to discard initial convictions when confronted with refuting evidence, something modern Christians still have to learn!

Though still referring to the Sun and other planets as souls, his next advance toward a truer explanation was his rejection of Aristotle's Intelligences or the scholastics' angels as the cause of their specific velocities, adopting the more realistic view that the reason their velocities decreased with their distance from the sun was due to the reduced influence of the sun. As he continues in the *Mysterium Cosmographicum*:

> But if...we wish to make an even more exact approach to the truth, and to hope for any regularity in the ratios [between the distances and the velocities of the planets], one of two conclusions must be reached: either (1) the moving souls are weaker the further they are from the sun; or (2) there is a single...soul in the center of all the spheres, that is...the Sun, and it impels each body more strongly in proportion to how near it is. In the most distant ones on account of remoteness and the weakening of its power, it becomes faint so to speak. Thus, just as the source of light is in the sun... so...the motion and the soul of the universe are assigned to that same Sun.... (p. 199; brackets added)

While still referring to the planets and the Sun as souls, his assertion of the Sun's "impelling power" to move the planets by an analogous force, that become less concentrated as they are dispersed, was another significant advance. Yet as an indication of how difficult it is to discard previous beliefs, he compared the Sun, Heavens, and Motive Force to God the Father, Christ the Son, and the Holy Spirit. (Cf. p. 63)

When the *Mysterium Cosmographicum* appeared in 1597 he sent copies to prominent astronomers, such as Maestlin, Tycho Brahe, and Galileo, the most favorable reply coming from Tycho who three years later offered him a position as his assistant in his observatory. In the interim, Tycho had accepted the position of Imperial Mathematicus extended by Emperor Rudolph II and, after moving to Benatky Castle near Prague, wrote Kepler in February 1599 asking to join him to collaborate in their astronomical research.

Searching for an exact mathematical correlation of the planetary orbits with their distances from the sun and learning that Tycho, owing to his exact naked eye astronomical observations had attained descriptions of the eccentric orbits of the planets, Kepler recognized that this was a most propitious opportunity. His first assignment was to measure the unusual eccentricity of the orbit of Mars realizing that it could be the key to determining the orbital parameters of the other planets. Bragging that he could resolve the problem in eight days, it took nearly eight years before he succeeded. Two and a half years later Tycho died ending their collaboration, but it proved advantageous for Kepler who succeeded him as Imperial Mathematicus to Rudolph II giving him continued access to Tycho's astronomical data and the financial support essential for his research.

Focusing on the eccentric orbit of Mars, he was struck by the fact that Mars' variations in velocity and brightness suggested that its orbit was egg shaped or ovoid and that at the perihelion where it is closest to the sun its speed increases, while at the aphelion it decreases contrary to the prevalent belief that the orbital motions were circular and uniform. Also, the analogy between the sun's emission of light and its propelling planetary force convinced him that rather than attributing this to the sun as a soul, it could be explained more realistically as a natural force (*vis*) emitted by the sun, another indication of his moving to a more scientific explanation. As he states in the *Mysterium*,

> once I believed that the cause which moves the planets was precisely a soul …but when I pondered that this moving cause grows weaker with distance, and that the Sun's light also grows thinner with distance from the Sun I concluded from this that this force is something corporeal…an emanation that a body emits, but an immaterial one. (p. 203, note 3)

In his comprehensive account of the great astronomers Arthur Koestler graphically describes the singular import of this change:

> It would be difficult to overestimate the revolutionary significance of this proposal. For the first time since antiquity, an attempt was made not only to *describe* heavenly motions in geometrical terms, but to assign them a *physical cause*. We have arrived at the point where astronomy and physics meet again, after a divorce which lasted for two thousand years. This reunion…led to Kepler's three Laws, the pillars on which Newton built the modern universe.[28]

But though he was ready to discard a spiritual cause, he was not yet prepared to adopt a purely physical one, describing the force as corporeal but "immaterial." Still, this indicates again a remarkable ability to revise his ideas when required by the evidence, showing that such cognitive changes are possible given a certain flexibility of mind.

In his major treatise, *Astronomia Nova*, he introduced a second force to explain the oval shape of Mars's orbit. Having read William Gilbert's book, abbreviated as *De Magnete*,[29] in which he described the earth and other planets (including the sun) as having magnetic poles on their opposite axes, Kepler surmised that as the planets revolved around the sun their respective polarities would be reversed causing them to be attracted or repelled depending upon the magnetic alignment of their poles. This magnetic force, added to the sun's emitting force, would explain their deviation from a circular orbit which could not be accounted for by the emitted force alone. Not only explaining the ovoid shape of the orbits, this also explained the variation in speed depending on the planets distance from the sun. What an extraordinary achievement!

With this additional force he was able to formulate two laws of motion as part of the final theory. Convinced that the emanating force of the sun decreases as the intervening spaces increase, he initially proposed an inexact inverse speed law, that the force from the Sun varies inversely with the distance, which he later corrected. Though mistaken, this enabled him to replace the ancient law of uniform circular motion by a second law stating that the "radius vector connecting the planet to the sun sweeps out equal areas in equal times," indicating that it is the *areas* not the *speeds* that vary uniformly with the time. Yet although these laws were based on the belief that the planetary orbits were ovoid, he still had not identified this shape as an ellipse.

He knew from his studies of Apollonius' treatise on conic sections that an ellipse was formed by a diagonal slice through a cone that resembled an ovoid shape, writing to Johann Fabricius that "if only the shape were a perfect ellipse all the answers could be found in Archimedes' and Apollonius' works."[30] It was only after measuring the exact deviation of the inner curve of Mars' orbit as .00429 of the radius that he concluded its shape was elliptical and thus arrived at his first astronomical law that the planetary orbits are elliptical with the sun as one of the foci. At last one could dispense with the ancient artifacts of circular orbits, crystalline spheres, epicycles, eccentrics, and equants with a simpler, more consistent, and more precise explanation! While these laws had been anticipated in his earlier *Commentaries on the Movement of Mars*, they were explicitly presented in his major work, *Astronomia Nova*, published in 1609 which has been acclaimed as "the first modern treatise on astronomy."

Having anticipated in his *Dioptrice*, published in 1611, the first part of Newton's universal law of gravitation, that two bodies attract proportional to their masses, he seems to have foreseen the second part that the force decreases with the *square* of the distance, not inversely or directly, as he previously had believed. For in that work he had claimed that the intensity of light decreases inversely with the square of the distance and since gravity was analogous to light, it would seem that this would apply to it also.

Then in the *Harmonice Mundi* published in 1619, despite his regressing to his earlier a priori approach and reconsidering the Pythagorean conception of the celestial orbs being orchestrated by a musical harmony that contradicted his empirical advances, when comparing the ratios of the periods of the planets with their distances from the sun based on Tycho's measurements, he nevertheless inferred the correct law from the fact that the squares of the periodic times are to each other as the cubes of the mean distances. As expressed today, the (periodic times)$^2 \propto$ (the mean distances from the sun)^3. Thus as implied by a statement in the *Harmonice Mundi*, the periods of the revolutions vary with the 3/2th power of their distances: "it is certain…*that the ratio which exists between the periodic times of any two planets is precisely the ratio of the 3/2th power of the mean distances, i.e., of the spheres themselves*…."[31] This was a crucial discovery because the ratio would provide Newton with the key to his universal law of gravitation.

Along with the exact laws, in the *Nova* he also introduced "gravity" as a mutually attractive force between physical objects analogous to magnetism, that is proportional to their masses regardless of their location in space. This was another critical concept in his formulation of a new astronomical framework, celestial mechanics!

> Gravity is the mutual...tendency between cognate bodies towards unity or contact... so that...[i]f two stones were placed anywhere in space near to each other, and outside the reach of force of a third cognate body, then they would come together, after the manner of magnetic bodies, at an intermediate point, each approaching the other in proportion to the other's mass.[32] (Brackets added)

This new physical force enabled him to explain the tides as caused by the combined gravitational forces of the Moon and the Earth. As he states in the Introduction to the *Astronomia Nova* : "if the Earth ceased to attract the water of the sea, the seas would rise and flow into the Moon...." (p. 338) (Unfortunately, Galileo will reject his explanation on the grounds that it involved occult forces acting at a distance.) Finally, he made the remarkably farsighted inference that the cosmos is a *heavenly machine that runs like clockwork,* a conception usually attributed first to Newton. As he clearly wrote to his friend Herwart regarding the intent of the *Nova* quoted by Koestler :

> "My aim is to show that the heavenly machine is not a kind of divine, live being, but a kind of clockwork...insofar as nearly all the manifold motions of the clock are caused by a simple weight. And I also show how these physical causes are to be given numerical and geometrical expression." (p. 340)

What an amazing progression of thought! He now rejects his earlier conception of the universe as a kind of spiritual entity consisting of souls for that of a mechanism, replacing the mysterious influence (analogous to light) that the soul-sun emanated to impel the planets with the physical forces of magnetism and gravity, along with declaring that the total system functions like clockwork that can be precisely described by his three laws of motion. This is all the more impressive considering that even William Gilbert, despite his experimental contributions to the understanding of magnetism and electricity, was unable to divest himself of the belief that the ultimate nature of magnetism was animistic, emanating from the magnetic soul.

Finally, in a work published in 1621 entitled *Epitome Astronomiæ Copernicanæ* in honor of Copernicus, which replaced the *Almagest* and the *De Revolutionibus* as the greatest astronomical treatises of the past, he presented a final generalization that went beyond the *Nova* in declaring that his three laws of motion applied to all the planets, not just to Mars, along with the Moon and the "satellites" of Jupiter, a term he introduced to designate the Moons of Jupiter which were telescopically discovered by Galileo. Yet there remained one last uncompleted project that he had intended to write for years, but never had found the time to devote to it, the *Rudolphine Tables*.

In 1614 John Napier had published a much praised work, *Merifici Logarithmorum Canonis Descriptio*, containing logarithmic tables that facilitated astronomical calculations, but had not shown how they had been computed. Knowing of its popularity but limited explication, in the years 1621-1622 Kepler wrote a book that contained not only logarithmic tables along with instructions for their use, but also considerable planetary data and a star catalogue comprising over a thousand fixed stars. Published five years later and entitled the *Tabula Rudophinæ* in honor of his deceased patron Rudolph II, it served for over a century as the basis of astronomical calculations and predictions.

Then, in addition to all his previous misfortunes, his death brought a final humiliation. In an attempt to recover 11,818 florins owed him by the Crown, he had gone to Ratisbon, the residence of the new Emperor, where after three days he fell ill and died alone on November 15, 1630, separated from his family. He was interred in the cemetery of Saint Peter outside of Ratisbon which later was so ravaged by numerous battles that the grave lacks a marker so that its exact location is unknown. If it were the case that there was a divinity guiding the destiny of human beings, it certainly would be disgraceful that the person who, by discovering the first three astronomical laws introduced a new cosmological system based on natural forces, a crucial advance in the evolving scientific method, should suffer such an ignominious end. This is partly mitigated by the fact that there now is a fine statue of Kepler and Tycho on the hill overlooking the city of Prague, but it still is a shameful injustice.

Unlike Copernicus whose astronomical approach was essentially conservative even if revolutionary in its consequences, Kepler advanced the provisional speculations of the scholastics toward the creation of the methodology of modern science by his discovery of mathematically exact astronomical laws and introduction of physical forces of explanation to replace spiritual powers. Despite these advances, because his accomplishments usually have been overshadowed by the more publicized attainments of Galileo and Newton, I have tried to redress this imbalance by stressing his remarkable intellectual development resulting finally in his complete break with the medieval period in his visionary conception of a mechanistic, clockwork universe.

Though astronomers considered him one of the leading scholars of the day, what also contributed to his lesser recognition was the fact that he was not attached to a major center of learning and that his astronomical laws were mathematical with as yet little empirical confirmation, aside from being based on the naked-eye observations of Tycho Brahe. Furthermore, that his discoveries were initially presented within a Neoplatonic perspective with considerable mystical overtones made them less convincing to some astronomers. Nevertheless, the title usually conferred on him as the "the father of modern astronomy" is completely justified.

The next contributor to the founding of modern science, William Gilbert, was older than Kepler by over a generation, but is being discussed later because unlike Kepler's association with Copernicus, Gilbert's research was entirely independent. Though his scientific contribution has been much less recognized than the other founders, his extensive scientific experiments in magnetism and electricity marked another significant advance in the formulation and application of the emerging scientific method. His great scientific work cited previously (f.n. 29), *De Magnete*, published in Latin in 1600 was not translated into English until 1893 by Fleury Mottelay, 293 years later, making it inaccessible to the general public. As late as 1836 Dr. John Davy complained that Gilbert's "'work is worthy of being studied, and I am surprised that an English edition (translation) of it has never been published.'" (p. xv) This explains why his research was not fully appreciated until the 19th century. Mottelay stated that translating the book has been "a task of no ordinary difficulty: it has brought up problems innumerable, the solution of which has involved much laborious research...." (p. v)

Gilbert was born in Colchester, Essex County, England, in 1540. According to the Biographical Memoir included in the Dover edition of *De Magnete*, little reliable knowledge is known of his early years, except that "he passed through the Grammar School of his native place and immediately afterward (May, 1558) entered St. John's College, Cambridge (whence, some say, he went to Oxford), proceeding B.A.," eventually earning an M.D. in 1569. (p. ix) After leaving St. John's he traveled on the Continent, "where probably he had the degree of Doctor of Physic conferred upon him, for he doth not appear to have taken it either at Oxford or Cambridge...." (p. x) He is said to have "'practiced as a physician with great success and applause'" and in 1573 "was elected a Fellow of the Royal College of Physicians, and filled therein many important offices, becoming...President (1600)." (p. x) Attracting the admiration of Queen Elisabeth I, "he was appointed her physician-in-ordinary" who also "showed him many marks of her favor, besides settling upon him an annual pension ...for the purpose of aiding him in the prosecution of his philosophical studies." (p. x)

These philosophical studies were first centered "almost exclusively on chemistry...but...was ere long made to yield to the study of the phenomena of electricity and magnetism, the latter of which had practically lain dormant for two thousand years—since the days of Thales and Theophrastus." (pp. x-xi) After eighteen years of exhaustive study of the extensive literature on magnetism and electricity and performing countless experiments, he published his great work with the long Latin title, *De Magnete magneticisque corporibus, et de magno magnete tellure; Physiologia nova, plurimis et arumentis et experimentis demonstrata*, fortunately abbreviated to *De Magnete*. The book eventually drew acclaim among scholars who finally recognized the originality and comprehensiveness of his research. His translator Mottelay declared that he was "'the father of the magnetic philosophy'" (p. v) while Joseph Priestly, the identifier of oxygen, claimed that he was "'the father of modern electricity.'" (p. xiii)

The following are further examples of his eventual acclaim. The eminent astronomer Sir J. F. W. Herschel in the eighteenth century declared that his book was "'full of valuable facts and experiments ingeniously reasoned on.'" (xi) The English historian Henry Hallam in his book, *Introd. to the Litt. of Europe in the 15th, 16th, and 17th Centuries* (London, 1839, Vol. II., p. 463), states:

> The year 1600...was the first in which England produced a remarkable work in Physical Science; but this was one sufficient to raise a lasting reputation for its author. Gilbert, a physician, in his Latin treatise on the Magnet not only collected all the knowledge which others had possessed on the subject, but became at once *the father of experimental philosophy in this island*, and, by a singular felicity and acuteness of genius, the founder of theories which have been revived after a lapse of ages, and are almost universally received into the creed of science.... (p. xii; italics added)

In his well-known *History of the Inductive Sciences* published in 1859 (Vol. II, p. 217), Dr. Whewell asserts that "'Gilbert's work contains all the fundamental facts of the science [of magnetism], so fully examined, indeed, that even at this day we have little to add to them....'" (p. xiii; brackets added) Thomas Thomson in his *History of the Royal Society* published in 1812 wrote that the *De Magnete* "'is one of the finest examples of inductive philosophy that has ever been presented to the world. It is the more remarkable because it preceded the *Novum Organum* of Bacon....'" (p. xiii) Indeed, Francis Bacon, Gilbert's contemporary, was one of his few detractors declaring in his *De Augmentis Scientiarum* that "'Gilbert has attempted to raise a general system upon the magnet, endeavoring to build a ship out of materials not sufficient to make the rowing-pins of a boat.'" (p. xv) As shall be pointed out later, Bacon raged against the limitations of scholastic philosophy and the status of general knowledge at the time, but was incapable of appreciating the significant scientific advances of such contemporaries as Gilbert, Kepler, and Galileo who, if he mentions them at all, does so derogatorily despite their acknowledged crucial scientific contributions.

> Bacon was ignorant of the fact reported in the *Biographical Memoir*, that Gilbert was the first to use the terms 'electric force,' 'electric emanations,' and 'electric attraction.' He it was also who gave the name of 'pole' to the extremities of the magnetic needle pointing to the poles of the earth, calling *south pole* the extremity that pointed toward the north,

and *north pole* the extremity pointing toward the south.... He considers the phenomena of electricity as having a considerable resemblance to those of magnetism, though he points out the differences by which the... phenomena are marked. (pp. xv-xvi)

Dr. John Robison, in his "System of Mechanical Philosophy" (London, 1822, p. 209) states:

> He has pursued...the subject of magnetism with wonderful ardour, and with equal genius and success; for Dr. Gilbert was possessed both of great ingenuity, and a mind fitted for general views of things. The work contains a prodigious number...of observations and experiments, collected with sagacity from the writings of others and instituted by himself with considerable expense and labour... [and] contains more real information than any writings of the age in which he lived and is scarcely exceeded by any that has appeared since. (pp. xvi-xvii; brackets added)

The concluding phrase, "scarcely exceeded by any that has appeared since," indicates that Dr. Robison, as the others just quoted, lived before the renewal of investigations into magnetism and electricity in the nineteenth century with remarkable success. It was then that Hans Christian Oersted (177-1851) discovered experimentally that a changing electric current produces a magnetic field and Michael Faraday (1791-1867) demonstrated that a changing magnetic field induces an electric current, their interdependence leading to the conception of electromagnetism as a field surrounding the charged body—rather than a fluid as was previously believed. Then James Clerk Maxwell (1831-1897) derived the equations describing the structure of the electromagnetic field and how it changes with time, inferring that since the velocity of an electromagnetic wave is the same as that of light, light too must be an electromagnetic phenomenon. Einstein described this latter identification as "one of the greatest achievements in the history of science."[33]

This brings us to Gilbert's book. Anyone who peruses *De Magnete* will be struck by the wealth of information it contains, both as to its exhaustive knowledge of pervious investigations as well as his own innumerable experiments exploring the various facets of magnetism in particular, but also of electricity, along with many diagrams and attractive woodcuts. He seems to have read every extract on magnetism and electricity by the ancient Greeks from Thales to Galen and all the research up to then, including those of such unlikely writers on the subject as Thomas Aquinas, Roger Bacon, and Petrus Peregrinus. Of Aquinas' "brief treatment of the loadstone," he asserted that he "gets at the nature of it fairly well" ((p. 7) and refers to the "small work attributed to Petrus Peregrinus " as "a pretty erudite book considering the time...." (p. 9) But usually he is scathing in his criticism of most past research, especially that of

Aristotle and the scholastics which then were held in such esteem. He wrote, for example, that regarding "the causes of magnetic movements, referred to in the schools of philosophers to the four elements and to prime qualities, these we leave for roaches and moths to prey upon." (p. 104) He declares that "[m]en of acute intelligence, without actual knowledge of facts, and in the absence of experiment, easily slip into error." (p. 82)

His own experimental discoveries and inventions were truly extensive and impresssive of which only a brief account of the main achievements can be presented here. After summarizing the previous research he describes the well-known properties of the lodestone and the magnet, noting that the "loadstone has from nature its two poles, a northern and a southern...which are the primary termini of the movements and effects" and that though the magnetic force varies in strength it is independent of the shape and "emanates from the parts themselves...the nearer they are to the poles of the stone the stronger virtues do they acquire and pour out on other bodies." (p. 23) The magnetized bodies attract metals and similar materials transmitting some of the "magnetic virtue" to the other body when their opposite poles attract, though not when they repel owing to a common polarity.

These polarities are acquired from the earth which, having a spherical shape, has its own north and south poles located at the opposite ends of its global axis. Using the earth as a model, Gilbert conducted many of his ingenious experiments "with the aid of a globe-shaped loadstone, as the best and most fitting," named the "Terrella" or "little earth." (pp. 23–24) As with the earth, one can draw meridians on the terrella by extending the pointed direction north of an iron needle into a circular line around the globe. Repeating this from different positions, one finds that the lines converge on the opposite poles forming a series of circumferences "creating meridians of longitude."

He then illustrates the principles underlying the function of the compass. Placing a lodestone with known north and south poles in a small wooden vessel on the surface of still water where it can float freely, he states that the stone "will straightway set itself, and the vessel containing it, in motion, and will turn in a circle till its south pole shall face north and its north pole, south." (p. 26) Apparently it was not then generally known that the opposite poles attract, the south pole of the compass pointing north and the north pole pointing south.

> Further, it is to be remembered that all who hitherto have written about the poles of the loadstone, all instrument-makers, and navigators, are egregiously mistaken in taking for the north pole of the loadstone the part of the stone that inclines to the north, and for the south pole the part that looks to the south: this we will hereafter prove to be an error. So illcultivated is the whole philosophy of the magnet still, even as regards its elementary principles. (p. 27)

His proof is presented in the form of a correct description of the alignment of the poles based on experiments with loadstones or charged pointer needles.

> The Loadstone moves and revolves until one of its poles, being impelled toward the north, comes to rest in its predetermined point on the horizon; the pole that comes to a stand looking north is (as appears from the foregoing…demonstrations) southern, not northern, though till now everyone has supposed it to be northern because it turns to the north. (p. 279)

However, determining the exact location of a position on the earth by the compass, such as on a ship at sea, also requires knowing the latitude, the distances north or south from the equator. The latter is the horizontal line that circles the earth exactly midway between the north and south poles, which represents 0° latitude and lies 90° from either pole. The other lines of latitude or parallels indicate the distances from the Equator, either north or south. Knowing the meridians or longitudes that mark the distances east or west on the earth's circumference, along with the latitudes north or south of the horizon line provide a grid for exactly locating any point on the surface of the earth. So for the compass to be useful there must be a way of determining the exact latitude of a ship's position.

> We come at last to that fine experiment, that wonderful movement of magnetic bodies as they dip beneath the horizon in virtue of their natural verticity…. This motion we have…illustrated and demonstrated with many experiments…so to point out the causes and reasons, that no one endowed with reason and intelligence may justly contemn, or refute, or dispute our chief magnetic principles. Direction, as also variation, is demonstrated on the plane of the horizon whenever a magnetic needle poised in equilibrium comes to a rest in any fixed point of it. (p. 275)

In Book V Gilbert meticulously describes the construction of an instrument that enables one to determine the exact latitude from the dip of the charged needle of a compass, along with a picture of a handsome "dip instrument." (p. 277) With it one discovers "that one of the needle's end which in northern latitudes looks to the north dips below the horizon; but in southern latitudes the end of the needle that looks south tends toward the earth's centre in a certain ratio…." (pp. 278-279) 'Magnetic dip' thus refers to the angle formed by a magnetic needle with the equator at any point due to the earth's magnetic field and polarization.

He was very aware of what a great difference the compass had made in navigating, especially in squalls or overcast skies when there are no identifiable landmarks or visible stellar guideposts so that the ships are driven aimlessly.

> A little ion bar—that soul of the mariner's compass, that wonderful director in sea voyages, that finger of God, so to speak—points the way and has made known the

whole circle of earth, unknown for so many ages. Spaniards (and Englishmen too) have again and again circumnavigated the whole globe on a vast circle by the help of the mariner's compass. (p. 223)

Along with these determinations he conducted electrical experiments with rubbed amber distinguishing its electrical attraction from the magnetic attraction of a lodestone. Using the electroscope or *versorium* that functions analogous to a compass when near electrically charged bodies, he describes discovering that opposite electrical charges attract while like charges repel, similar to magnetism. (Cf. pp. 79-97) He also discusses Copernicus's heliocentric theory, agreeing that the two motions he attributed to the earth were simpler and more harmonious than the explanations of either Aristotle or Ptolemy and recognizing the central role of the sun in producing solar motion.

> The earth therefore rotates, and by a certain law of necessity, and by an energy that is innate, manifest, and conspicuous revolves in a circle toward the sun; through this motion it shares in the solar energies and influences; and its verticity holds it in this motion lest it stray into every region of the sky. The sun (chief inciter of action in nature), as he causes the planets to advance in their courses, so, too, doth bring about this revolution of the globe by sending forth the energies of his spheres – his light being effused. (p. 333)

As usual, his most scornful criticism of the opponents of heliocentrism is directed toward Aristotle, especially his Prime Mover: "This *primum mobile* presents no visible body, is in no wise recognizable; it is a fiction believed in by some philosophers, and accepted by weaklings who wonder more at this terrestrial mass here than at those distant mighty bodies that baffle our comprehension." (p. 321) He also rejects Aristotle's conception of the celestial orbs or spheres, expressing an exceptional recognition of the vastness of the universe, declaring that in addition to the visible stars there must be many more that are unseen: "What, then, is the inconceivably great space between us and these remotest fix stars? and what is the vast immeasurable amplitude and height of the imaginary sphere in which they are supposed to be set?...they are beyond the reach of eye, or man's devices, or man's thought. What an absurdity is this motion (of spheres)." (pp. 319-320) How pleased he would have been at Galileo's telescopic discoveries that greatly extended man's vision and understanding of the solar system.

I have focused on the positive achievements of Gilbert and how these were commended centuries later to redress the fact that he is one of the least appreciated scientists of that era, decidedly more than Kepler. But to be fair, it should also be pointed out that much in the book, such as his analysis of the nature of metals, reads like science fiction because of the limited knowledge at the time. This is particularly true of his explanation of the nature of magnetism which he regarded as animistic owing to the celestial bodies possessing souls. Even someone as empirically oriented as he, who put so much trust in experimental tests, could not divest himself of the ancient view that only souls could produce the diverse motions and harmonious order of the universe. This is especially surprising considering his scornful criticism of Aristotle's Prime Mover and spheres. Yet he states:

> Wonderful is the loadstone shown in many experiments to be…as it as were, animate. And this one eminent property is the same which the ancients held to be a soul in the heavens, in the globes, and in the stars in sun and moon. For they deemed that not without a divine and animate nature could movements so diverse be produced, such vast bodies revolve in fixed times, or potencies so wonderful be infused into other bodies; whereby the whole world blooms with most beautiful diversity through this primary form of the globes themselves. (p. 308)

Lest there be any doubt about his own conviction in this regard, he claims: "As for us, we deem the whole world animate, and all globes, all stars, and this glorious earth, too, we hold to be from the beginning by their own destinate souls governed and from them also to have the impulse of self-preservation." (p. 309) Unlike Kepler, his younger contemporary, he was unable to divest himself of the primitive view that only spirits, souls, or divine beings could be the source of motion. But even today creationists believe that the order in the universe attests to a divine being rather than natural causes and laws. How long will it take to rid ourselves of the illusion that the concept of the soul or divine spirit explains anything, whether human consciousness or the movement of the stellar bodies?

CHAPTER SIX

Galileo

In contrast to the delayed and rather limited recognition respectively of the contributions of Gilbert and Kepler, Galileo's teaching positions at the Universities of Pisa and Padua, his patronage by the Medici, his renown controversial telescopic observations, and that his major works were written in the more accessible Italian and dialogue form brought him worldwide recognition, though somewhat later in life. In addition, his demolishing of the Aristotelian cosmological system and replacement of the latter's conception of scientific explanation by a precise formulation of the methodology of the new science incorporating mathematics, his experimental proof of the law of gravitational fall, and his disgraceful condemnation by Pope Urban VIII and the Inquisition because of the preference he showed for the heliocentric system in his major work, *Dialogue Concerning the Two Chief World Systems*, assured his everlasting fame. As many books have been written about his background and life, I shall focus on his contributions to the transition from the Renaissance period of scientific inquiry to the creation of modern classical science in the 16th and 17th centuries.

Galileo enrolled as a medical student at the University of Pisa in 1581 at his father's insistence that he renew the family's distinguished heritage in medicine derived from Galileo Bonaiuti, a prominent physician of the 15th century, but after overhearing a lecture on Euclid by the court mathematician Ostilio Ricci he was captivated by mathematics. Introduced to the works of Eudoxus and Archimedes by Ricci who recognized his unusual mathematical aptitude, this led to his early studies and research in statics and hydrostatics. His first scientific publication, *La Bilancetta* (*The Little Balance*), appeared in 1586, a book in Italian proposing modifications in the Westphal balance to improve measurements of specific gravities and refining Archimedes' method for calculating specific weights. He began writing a dialogue on motion that he never finished but that was the preparation for his later treatise, *De Motu*, published in 1590.

He had acquired enough of a reputation that when the chair of mathematics at the University of Pisa was available in 1589 he obtained the position, even though he was only twenty-five and had not acquired a university degree. During his three-year appointment his research consisted mainly in criticizing Aristotle's distinctions and explanations of motions that were later included in the *De Motu*. Then, despite his lack of publications, his reputation was sufficient to compete successfully for the chair in mathematics at the University of Padua outside of Venice. This had many attractions since Venice was not only a uniquely beautiful cosmopolitan city, it was an artistic, artisanal, and munitions center as well as a great seaport visited by many well-known scholars. The influence of Venice can be seen in the numerous references in his books to sailing ships, compasses, sextants, pulleys, levers, and cannons. Extraordinarily curious, his inquiries extended to machines, incline planes, magnets, and optics, along along with his telescopic observations. Skilled at using and inventing instruments, such as the telescope, microscope, calculator, quadrant, and various calibrators, he hired an instrument maker to live with his family to help in building his inventions.

His first years at the University of Padua were devoted to mechanistic inquiries, although this was gradually replaced by a growing interest in Copernicus' theory since he believed that the two motions attributed to the earth by Copernicus could explain the tides. He knew of Kepler's theory that they were caused by the mutual gravitational forces between the earth and the moon, but rejected it because it seemed to invoke mysterious occult forces acting at a distance. Instead, drawing on the analogy of the to and fro motion of the bilgewater in sailing ships as they plowed through rough seas, he believed that the two motions attributed to the earth by Copernicus produced the tides. Although opposed to Kepler's explanation of the tides in terms of gravitational forces, in a letter acknowledging his gift of the *Mysterium Cosmographicum*, Galileo added that he rejoiced to see that Kepler supported Copernicanism, declaring

> from that position I have discovered the causes [an allusion to the earth's motions?] of many physical effects [the tides?] which are perhaps inexplicable on the common [geocentric] hypothesis. I have written many...refutations of contrary arguments [to the Copernican system] which up to now I I have preferred not to publish,

intimidated by the fortune of our teacher Copernicus, who though he will be of immortal fame to some, is yet by an infinite number (for such is the multitude of fools) laughed at and rejccted.[34] (Brackets added)

Reappointed by the University for four more years, he completed a book on *Mechanics* in 1602 that contained a renewal of his incline plane experiments described in the *De Motu*, along with his discovery of the law of free fall. He also measured the ratio of the orbital sizes and speeds of the planets in relation to their distances from the sun, but unlike Kepler was unable to calculate the exact proportion. Yet experimenting with pendulums he made the major discovery that when the duration of their swings were equal or isochronal it was the length of the pendulum, not the weight of the body, as believed, that determined their speeds. This was pivotal because it suggested that in a vacuum the velocity of falling objects would be the same regardless of their weights and also illustrates the importance of experiments in correcting mistaken beliefs.

Continuing the incline plane investigations, two years later he experimentally confirmed that accelerating from rest along an inclined plane the distances traversed by a smooth round object in equal times were as the odd numbers beginning with one (1, 3, 5, 7, 9, &). He then noticed that the square roots of the *successive sums* of the odd numbers gave the times (2, 3, 4, 5 &) that, when squared, showed the ratios of the increases in acceleration. Following this discovery, he wrote to Palo Sarpi that "'he had found a proof for the square law, the odd-number rule, and other things he had long been asserting, if granted the assumption that velocità are proportional to distances from rest.'" (p. 100) In the *Two New Sciences*, his final book written shortly before his death, this law would be the key to the new science of falling objects.

In 1604 a brilliant nova was sighted arousing controversy as to its location in the sky. Then in the summer of 1609, when he was forty-five years old, Galileo learned of a Dutch instrument that magnified distant objects causing them to appear closer, initially called a "spyglass," arousing his usual curiosity in novel instruments. As he wrote in the *Sidereus Nuncius or The Sidereal Messenger* (*Starry Messenger*):

> This…caused me to apply myself totally to investigating the principles and figuring out the means by which I might arrive at the invention of a similar instrument which I achieved shortly afterward on the basis of the science of refraction. And first I prepared a lead tube in whose ends I fitted two glasses, both plane on one side while the other side of one was spherically convex and the other concave. Then, applying one eye to the concave glass, I saw objects…three times closer and nine times larger than when observed with natural vision only. Afterward I made another more perfect one for myself that showed objects more than sixty times larger.[35]

Adding that sparing no labor or expense "he constructed...an instrument so excellent that things seen through it appeared about a thousand times larger and more than thirty times closer than when observed with the natural faculty only." (pp. 37–38) He demonstrated its power to the Venetian Senate from the campanile, the highest point in St. Mark's Square in Venice, and presented the Doge with a gift of the spyglass which brought him a lifetime appointment at the University of Padua. His first lunar observations were made in December 1609 (the same year Kepler published his *Astronomia Nova*) with drawings of the crescent shape of the Moon and its irregular surface followed by the startling discovery of the "four little stars" circling Jupiter.

The latter was startling because in Aristotle's cosmology the inherent perfection of the celestial realm precluded any changes or new discoveries in the translunar world. Galileo published his observations in the *Sidereus Nuncius* the following March which immediately brought him international acclaim. His diagrams of the surface of the Moon, particularly the mountainous terrain and steep valleys resembling those on the earth, were especially shocking because they too challenged Aristotle's conception of the planets as "ethereal bodies" and distinction between the celestial and terrestrial worlds.

> By oft-repeated observations...we have been led to the conclusion that we certainly see the surface of the Moon to be not smooth, even, and perfectly spherical, as the great crowd of philosophers have believed about this and other heavenly bodies, but on the contrary to be uneven, rough, and crowded with depressions and bulges. And it is like the face of the Earth itself, which is marked here and there with chains of mountains and depths of valley. (p. 40)

The threat to Aristotle's cosmology was so grave that its defenders devised desperate counter arguments, such as claiming that the Moon was surrounded by a transparent crystalline sphere covering its irregular surface or that the spyglass was only reliable when directed to the terrestrial, not the celestial world. The influence of authority was so strong that the Aristotelian Cremonini argued that such phenomena could not exist because they were not mentioned by Aristotle, refusing to look through the spyglass. In fact it did take practice and skill to use the instrument effectively, but for Galileo the choice between accepting his new telescopic observations or the ancient views of Aristotle was obvious.

Owing to his recent discoveries, he was offered the position of chief Mathematician at the University of Pisa and Philosopher to the Grand Duke of Tuscany which he readily accepted despite his recent lifetime appointment to the University of Padua. It was more appealing because there actually were no teaching obligations, despite the title, and Florence was even more attractive to him than Venice because it was his original home and provided the patronage of the powerful

Grand Duke Cosimo II. As an indication of the latter's support "the Tuscan ambassadors at the courts in Prague, London, Paris, and Madrid were alerted that Galileo would send them copies of his book and perhaps spyglasses as well," instructing them "to use their good offices to promote Galileo's discoveries." (p. 100) By the end of the year his observations of the "moons of Jupiter," as they were now called (and would shortly be renamed "satellites" by Kepler), had been confirmed by Antonio Santini in Venice, Kepler in Prague, and Thomas Harriot in England, among others.

Continuing his observations, he discovered what are now called the rings of Saturn but which he described as two opposite protuberances resembling ears, as well as the phases of Venus. The latter were particularly significant because they were predicted by Copernicus from the heliocentric perspective, while from the Aristotelian or Ptolemaic systems they would have presented a constant crescent shape of different sizes, thus offering further confirmation of heliocentrism. His discoveries having now gained more credibility, he went to Rome in March 1611 where the authenticity of his sightings were certified by the Collegio Romano (the prestigious Jesuit School) before Cardinal Bellarmine (who will play such an important role at his later trial). It was at a banquet in his honor, after being inducted into the prestigious Accademia dei Lyncei, that the name *telescopium* was introduced to replace the previous *occhilai* or spyglass.

Yet the controversies that stalked him continued. In a disagreement with the Aristotelians over floating bodies which they erroneously attributed to the object's shape, Galileo defended Archimedes' principle of specific gravity, that it was the ratio of the objects' density to water that explained its support, the rebuttal later published (in translation) as a *Discourse on Bodies On or In Water*. Then in 1612 he entered into a further dispute with a Jesuit mathematician named Christopher Scheiner who, at the request of his order wrote under the pseudonym Apelles, claimed that the recently detected sun spots were tiny stars similar to Jupiter's Moons revolving around the sun's surface. In his three replying letters, Galileo argued that his astronomical observations indicated they were not stars but more like the clouds circling the earth.

In the third letter he expresses his early skepticism (which he later will modify) regarding the prospects of scientific explanations, similar to the positivists' position during the first two-thirds of the twentieth century. On that view scientists were incapable of discovering the more basic structures or causes of phenomena and therefore should limit explanations to observed correlations in nature. As stated by Galileo:

> For in our speculating we either seek to penetrate the true and internal essence of natural substances or content ourselves with a knowledge of some of their properties. The former I hold to be as impossible…with regard to the closest elemental substances

as with more remote celestial things.... I know no more about the true essences of earth or fire than about those of the moon or sun, for that knowledge is withheld from us.... But if what we wish...is the apprehension of some properties of things, then it seems to me that we need not despair...to acquire this respecting distant bodies just as well as those close at hand—and...in some cases even more precisely in the former than in the latter.[36]

While this restricted view of scientific discoveries being limited to the superficial properties of things was justified in Galileo's day, it was not reasonable by the time of the positivists. Since the end of the 19th century it had been refuted by the theory of electromagnetism, the discovery of subatomic particles such as electrons, protons, neutrons, and photons, along with alpha, beta and gamma rays, the determination of the atomic-molecular structure of substances and elements, the creation of atomic fission, and more recently the detection of the microwave background radiation left over from the Big Bang. It is possible the universe is so vast or complex that no final theory is attainable, yet that does not preclude the entities previously discovered being real within the particular conditions in which they exist and are detected.[37] Electrification, discovery of extragalactic universes, explanation of chemical and biomolecular structures and reactions, creation of the atomic bomb, landing on the moon, and decoding the genome would not have been possible if our theoretical explanations were as limited as the positivist's claimed.

But Galileo's argument, as those of Gilbert, was directed against the scholastics who claimed that all knowledge could be found in the writings of Aristotle, "as if this great book of the universe had been written to be read by nobody but Aristotle, and his eyes had been destined to see all for posterity." (p. 200) He added that were Aristotle still alive and given his respect for empirical evidence, it is unlikely he would oppose the recent discoveries. He even had the audacity (for that time) to declare, in a famous Letter to Castelli on December 13, 1613, that concerning salvation and the establishment of the Faith there was no higher authority than Holy Scripture, but

> I should think it would be prudent if no one were permitted to oblige scripture... to sustain as true some physical conclusions of which sense and demonstrative and necessary reasons may show the contrary," especially as it seems unlikely that the same God who has given us our senses, reason, and intelligence wished us to abandon their use.... (p. 226)

Given the fact that in America today a vast number of Christians reject the overwhelming empirical evidence for evolution and the Big Bang explanation of the origin of the universe because of their religious beliefs, Galileo's admonition is as valid today as it was then. At the time he was throwing down the gauntlet to the Catholic Church claiming that questions involving sensory observation and rational demonstration were the provenance of natural philosophers (i.e., scientists), not ecclesiastical authorities. A year later on December 21, 1614 the issue exploded in Florence when a fiery young Dominican, Tommaso Caccini, "denounced from the pulpit of Santa Maria Novella the Galileists, and all mathematicians along with them, as practitioners of diabolical arts and enemies of true religion." (p. 238)

While in fairness it should be said that in Florence and Rome many Dominicans disavowed Caccini's vindictive remarks, the dispute initiated the proceedings that eventuated in Galileo's being brought to trial and convicted of heresy by the Inquisition. For following Caccini's attack, the cardinals of the Inquisition began examining Galileo's Letter to determine whether it contained offensive material forcing him toward the end of 1615 to journey to Rome to allay any charges of heresy and to persuade the Church authorities not to proscribe discussion of the Copernican theory. But unlike his successful visit to Rome four years earlier, when he was honored by the members of the Collegio Romano, this undertaking proved distinctly unfavorable as indicated by Drake's description of the severity of the proceedings.

> First a technically independent panel of theologians would find against the notions that the earth moved and the sun stood still, and then Galileo would be informed of this and asked to abandon those views. Bellarmine had no doubt that he would agree; a decree of general scope could then be published and the matter would be resolved without alienating either the Medici or the several cardinals who remained favorable to Galileo. The finding of the panel was handed in on 24 February [1616]; on the 25th, at the weekly meeting of the cardinals of the Inquisition, the pope [Paul V] instructed Bellarmine in their presence to inform Galileo of it and require him to abandon these opinions. If he resisted, then the commissary of the inquisition was to instruct Galileo in the presence of a notary and witnesses that if he did not obey he would be jailed. (p. 253; brackets added)

As Bellarmine counseled, on February 26 before a notary, witnesses, and other officials he met with Galileo. A notarized record of the meeting, though unsigned, states that "he told Galileo of the official findings against the motion of the earth and stability of the sun," while the commission of the Inquisition in his presence "admonished Galileo in the name of the pope that he must not hold, defend, or teach in any way, orally or in writing, the said propositions on pain of imprisonment. Galileo Agreed." (p. 253)

The importance in relation to the later trial and conviction is that Galileo *did agree* "not to hold, defend, or teach in any way, orally or in writing," the official findings against the motion of the earth and stability of the sun. An edict was then issued proscribing all books intending to reconcile heliocentrism with Christian doctrine, although none of Galileo's books were included. Having previously conceded that the Church had the authority to decide theological questions, it was whether they had an equal authority to decide *questions of natural philosophy* that Galileo was contesting!

Concerned that these proceedings would be distorted by his adversaries, Galileo asked to speak to the pope directly. Given the pope's previous stand, he was surprised at his cordial reception, assuring him that "he knew Galileo's integrity and sincerity, told him not to worry, said that not only he but the entire Congregation of the Holy Office knew about his unjust persecutions, and added the unusual remark that so long as he, Paul V lived, Galileo remained secure." (p. 256) Learning that it was being rumored that he had been severely admonished and forced to do penance, when he informed Bellarmine who previously had shown him consideration, the latter "wrote out a signed statement that Galileo had neither adjured nor done penance, but had merely been informed of the general edict governing all Catholics." (p. 256)

Then in a further dispute in the fall of 1618 regarding the location of three just sighted alleged comets, Orazio Grassi, another professor at the Collegio Romano, in a public lecture supported Tycho Brahe's previous arguments that because there was no evidence of parallax or enlargement and they were seen close to Venus, they must exist in the translunar world. While Galileo normally would not have been hostile to this interpretation, because Grassi's arguments depended on adopting Tycho Brahe's modified geocentric system in which the sun, encircled by the planets, revolved around the earth, a view Galileo found repugnant, he felt compelled to respond despite the former prohibitions from engaging in such topics. Being ill at the time, his reply was presented by his assistant Mario Guiducci to the Florentine Academy, though there is no question as to who wrote it.

In an unduly derisive rebuttal, he argued that if the observed phenomena were not comets neither the lack of parallax nor of magnification would establish the nature of their orbits. He even mocked the question as to whether they were in the translunar or sublunar world, "[n]ever having given any place in my thoughts to the vain distinction (or rather contradiction) between the elements and the heavens...."[38] Thus he succeeded not only in antagonizing Grassi, but also in aroused the hostility of some Jesuits at the Collegio Romano who had previously honored him by authenticating his astronomical observations. Grassi's reply in the *Libra Astronomia* or *Astronomical Balance*, the title indicating that he was carefully weighing Galileo's arguments, led to a second critical rebuttal by Galileo entitled *Il Saggiatore* or *The Assayer*, implying that he in turn was meticulously evaluating Grassi's reasoning, although he had to admit that "I cannot determine precisely the manner in which comets are produced...knowing that it may occur in some way far beyond our power to imagine." (p. 236)

But the significance of this later work lies in a fuller description of his theory of knowledge and of scientific inquiry than previously presented. Like the ancient Atomists but contrary to Aristotle, Galileo held that we do not perceive the world as it is, but as it appears to us due to the human sensory system, the sensorium transforming unobservable physical stimuli, such as light or sound waves that stimulate our sense organs, into the experienced world of configured colored objects and sounds, as well as sensations like smell, taste, heat, and pain. Most of the major contributors to modern epistemology, such as Descartes, Locke, Hume, and Kant were dualists, although none that I know of indicated they were influenced by Galileo.

Descartes was a contemporary of Galileo, but he was so critical of his experimental claims and had such a different conception of scientific inquiry that it is unlikely that Galileo influenced him, while Locke was a contemporary of Newton and Hume and Kant were of the eighteenth century. Locke was a close friend of Newton and thus his well-known distinction between the independently existing physical properties of objects, that he called primary qualities in contrast to the subjective sensory qualities that he designated secondary qualities, could have been suggested by Newton's adoption of the corpuscular theory as the underlying ontology of his scientific framework in contrast to the world of experience. In any case, Galileo presents one of the earliest modern statements of scientific dualism in his *Il Saggiatore (The Assayer)*.

> Therefore, I say that upon conceiving of a material or corporeal substance, I immediately feel the need to conceive simultaneously that it is bounded and has this or that shape; that it is in this place or that at any given time; that it moves or stays still.... I cannot separate it from these conditions [its primary qualities] by any stretch of my imagination. But that it must be white or red, bitter or sweet, noisy or silent, of sweet or foul odor [the secondary qualities], my mind feels no compulsion to understand as necessary accompaniments.... For that reason I think that tastes, odors, colors, and so forth are no more than mere names so far as pertains to the subject [object] wherein they [appear to] reside, and they have their habitation only in the sensorium. Thus, if the living creature (l'animale) were removed, all these qualities would be removed and annihilated. (p. 309; brackets added)

Since according to Aristotle our perceptions normally reproduce the forms and the qualities of things as they exist independently in them, while Galileo claimed that only the primary qualities of objects could exist in them, this was another contentious difference with the Aristotelians, one that posed one of the crucial problems in modern philosophy, Kant arguing that *both* primary and secondary qualities are imposed by the mind, things in themselves being inexperienced and unknowable.

In addition to his dualism, *Il Saggiatore* presents his famous declaration of the significance of mathematics in scientific inquiry, as Grosseteste had done, rejecting Aristotle's syllogistic logic as the primary explanatory formalism of science.

> Philosophy is written in this grand book—I mean the universe—which stands continually open to our gaze, but it cannot be understood unless one first learns to comprehend the language and interpret the characters in which it is written. It is written in the language of mathematics, and its characters are triangles, circles, and other geometrical figures, without which it is humanly impossible to understand a single word of it; without these, one is wandering about in a dark labyrinth. (pp. 183–184)

Despite limiting the characters of mathematics to geometrical figures without mentioning algebraic notation and equations, this is one of the most eloquent statements (following Grosseteste) ever written regarding the significance of mathematics for scientific inquiry that would have tremendous impact.

Pope Paul V having been succeeded by Pope Gregory XV who in turn was succeed by Pope Urban VIII, Galileo believed that he no longer had to fear oppression by the Church because Urban had been such an early admirer of him that he had even written a poem in his honor. In reciprocation the Lincean Academy, which had sponsored publication of the *Il Saggiatore*, had it dedicated to the pope with his armorial crest of three bees printed on the cover. It so pleased him that he had portions read to him at table. In these improved circumstances, Galileo considered it propitious for taking another journey to Rome to see if he could get permission to write the book he had long intended on the comparative evidence for the Aristotelian-geocentric and Copernican-heliocentric systems, despite the previous prohibitions.

Succeeding in having six audiences with the pope in April 1624, he was given permission to write the book with the pope even reassuring him (in Drake's words) "that the Holy Church had never condemned Copernicus as heretical, and never would, holding it only to be rash, there being no fear that anyone would prove it true" (p. 291), an indication that the pope was unaware of the proceedings of 1616 prohibiting discussion of the Copernican system and further evidence of how little confidence there was at the time of resolving these cosmological questions.

Though the meeting was reassuring, there began the lamentable consequences based on inadvertent omissions and misunderstandings. Galileo either unintentionally or deliberately neglecting to tell the pope of the various proceedings of February 1616 in which "he had been admonished…that he must not hold, defend, or teach, in any way the official findings against the motion of the earth and the stability of the sun." When Urban was informed of this he became convinced that

Galileo had deliberately deceived him by withholding the information. As related by Drake, Galileo

> left Rome with assurances from Urban that he was free to write on the two systems of the world provided that he treated them impartially and not go beyond the astronomical and mathematical arguments.... But because Bellarmine had instructed Galileo in 1616 to regard the admonition given him by the Commissary of the Inquisition as having no official existence, so long as Bellarmine's own words to him were heeded, Galileo never told Urban VIII what had actually taken place on that occasion. In the end the omission on his part turned out to have been a fatal error. (p. 291)

On his return from Rome Galileo learned of a critical appraisal of Copernicus's system written earlier by Francesco Ingoli that he took as an opportunity to renew the controversy by writing *Reply to Ingoli* in 1624. The *Reply* is of special interest because it predates two of Galileo's arguments in *Dialogue Concerning the Two Chief World Systems*: the first dismissing the distinction between the terrestrial and celestial worlds as being merely verbal and the second a refutation of arguments allegedly proving the earth could not move. Regarding the first, Ingoli had argued that as the heaviest body the earth naturally exists in the central or lowest position of the spherical cosmos while the weightless, unchanging aither naturally dwells in the highest region, distinguishing the sublunar and translunar worlds. Galileo again rejected this designation as merely verbal, stating "that those ['lowest' and 'highest'] are words and names, proving nothing and having nothing to do with calling anything into existence." (p. 293; brackets added)

Ingoli's second argument declares that if the earth moved, then objects projected vertically upward would not fall to their original position, as they obviously do, because during the time of their ascent and descent the earth would have rotated so that the objects would fall some distance away from its launching spot, its distance and direction depending on the speed of the earth's rotation, as well as the height of the ascent. To illustrate, the Aristotelians claimed that a solid object dropped from a tall masthead of a swiftly moving sailing ship would fall some distance behind the base of the masthead, while if dropped from a stationary ship it would fall parallel to the masthead.

As with most arguments of this type, although seemingly reasonable to our ordinary intuitions, they overlook that fact that an object falling on the moving ship *partakes of two motions*, the vertical descent plus the motion conveyed to the object by the system within which it is moving, the earth or the ship, with the latter canceling out. For example, within the cabin of a *uniformly* moving ship one cannot tell, within the ship, whether it is in motion or not which is true of any uniformly moving system, such as an airplane, the laws of inertial motion being the same as those at rest.

Descartes and Mersenne declared that already being convinced of the falsity of the Aristotelian argument Galileo did not perform the experiment, but with an open mind it is difficult to doubt that what he wrote, as quoted by Drake, was true: "I have been twice as good a philosopher as those others because they, in saying what is the opposite of the effect, have also added the lie of their having seen this by experiment; and I have made the experiment—before which, physical reasoning had persuaded me that the effect must turn out as it indeed does." (p. 294) Later we shall find that he was also falsely accused of not having performed the incline plane experiments because he admitted he had deduced the results in advance. Yet Galileo was too much of an experimentalist not to realize the importance of testing one's deductions experimentally, as his precise description of the experiment indicates, and his integrity would not have permitted him to lie about having made them.

A year following his *Reply to Ingoli* he began writing his famous *Dialogue Concerning the Two Chief World Systems*, but due to ill health and other projects it was not finished until the end of 1629; then, because of difficulties obtaining the imprimaturs related to the earlier prohibitions, it was not published until 1632. In my opinion the greatest polemical masterpiece in the history of science, the *Dialogue* establishes Galileo, along with Plato, as the two master dialecticians of the ages. His deconstruction of the underlying assumptions, principles, concepts, and definitions of the Aristotelian system that had generally dominated natural philosophy since Thomas Aquinas's great synthesis in the 13th century, replacing it with the essential methodology and preliminary theoretical framework of modern science, was one of the prime intellectual achievements of humankind, although its complete fulfillment would await Newton. But the avowal of this achievement is not meant to denigrate Aristotle's greatness since his grand synthesis of knowledge also was a crucial stage in man's effort to free himself from mysticism and mythology to attain a more realistic understanding of the universe.

The dialogue occurs over four days among three interlocutors, two of whom had been close friends and supporters of Galileo who died before the book was published and whose memory he wanted to commemorate. One was Salviati, a Florentine aristocrat presented as the Academician who represents Galileo, the other Sagredo, a Venetian nobleman who is the sagacious moderator in the dialogues. Simplicius, the third disputant, is named for a sixth century scholastic who defends the philosophical system of Aristotle.

On the first day the dialogue discusses the distinction rejected by Galileo in his *Reply to Ingoli* between the sublunar and the translunar worlds which, if true, would preclude the terrestrial earth being in the celestial world. As previously, Galileo has Salviati argue that the distinction is unwarranted based on definitions which do not have the same conceptual significance, given new evidence, as they did in Aristotle's day, such as "perfect circular versus imperfect rectilinear motions," "natural as opposed to unnatural or violent motions," and the "ethereal, weightless, incorruptible celestial realm" in contrast to the "material, dense, and corruptible terrestrial world." As evidence he cites the recently observed *rectilinear* motion of meteors observed in the translunar realm and instances of *circular* motion in the sublunar to show that the difference based on the two kinds of motion is artificial.

To rebut the qualitative distinction between the two realms he refers to his telescopic observations of the similarity between the surface of the Moon and the Earth and the resemblance between the spots circulating the sun and the clouds drifting above the earth. To reject the perfection of the celestial realm based on its immutability he cites the recent discovery of meteors, novas, and new stars as evidence of changes in the translunar world. However, these arguments do not persuade Simplicius who replies:

> This way of philosophizing tends to subvert all natural philosophy, and to disorder and set in confusion heaven and earth and the whole universe. However, I believe the fundamental principles of the Peripatetics [followers of Aristotle] to be such that there is no danger of new sciences being erected upon their ruins."[39] (Brackets added]

Before turning to the main dialogue of the second day, there is a diversionary discussion with Simplicius maintaining, similar to the early Church Father's defense of scripture, the complete authority of Aristotle's writings: "There is no doubt that whoever has the skill will be able to draw from his books demonstrations of all that can be known; for every single thing is in them." (p. 108) While today most people would be shocked by this statement given all the advances in the various sciences since then, we should remember how impressive the works of Aristotle must have appeared at that time given the lack of empirical inquiry during the medieval period and how conditioned the scholastics were to accepting the authority of the Catholic Church. But Galileo had already presented his rebuttal at the end of the previous day's dialogue when Sagredo tersely declared "there is not a single effect in nature, even the least that exists, such that the most ingenious theorists can arrive at a complete understanding of it. This vain presumption of understanding everything can have no other basis than never understand anything." (p. 101)

Turning then to the main topic of the day, the arguments for the diurnal eastward rotation of the earth to account for the apparent westward movement of the sun, Salviati argues that because of the relativity of uniform motions the question must be decided on more than the appearances. Just as it seems from a smoothly moving ship that the coastline is receding while from the shore it is the ship that is withdrawing, which is the correct view cannot be settled by appearances alone, but by some additional evidence or reference point. Thus whether it is the eastward rotation of the earth or the westward revolution of the fixed stars carrying the sun cannot be determined merely by looking and since there is no fixed reference to settle the issue, it has to be decided on which is the simpler, more harmonious hypothesis, as Salviati argues.

> First, let us consider only the immense bulk of the starry sphere in contrast with the smallness of the terrestrial globe.... Now if we think of the velocity of motion required to make a complete rotation in a single day and night, I cannot persuade myself that anyone could be found who would think it the more reasonable and credible thing that it was the celestial sphere which did the turning, and the terrestrial globe that remained fixed. (p. 115)

Because the argument depends upon what is considered to be "the more reasonable and credible thing," one could reply that that too is relative since it rests on whose consideration is chosen. But additional arguments follow.

As for the criterion of harmony, consider the discord resulting from the fixed stars located in the furthest sphere rotating daily in a brief twenty-four hours while the orbital speeds of the planets decrease with their distances from the center, Saturn requiring thirty years to circle its orbit in contrast to the stars' twenty-four hours. Other discrepancies follow from assuming the earth to be stationary, such as the contrasting diurnal westward rotation of the whole universe while each of the planets revolve in their orbits in the contrary direction eastward. But, Salviati concludes, "by making the earth itself move, the contrariety of motions is removed, and the single motion from west to east accommodates all the observations and satisfies them all completely." (p. 117)

He next reconsiders the argument discussed in *Reply to Ingoli* that if the earth rotated from west to east then objects in the air, whether projectiles, birds, or clouds, would be seen to move westward by someone standing on the earth. Repeating Galileo's previous argument in the *Reply*, Salviati points out that in the cabin of a ship moving uniformly all moving objects, dropped or thrown, behave as they would if the ship were at rest, just as he had demonstrated that a solid object dropped from the masthead fell parallel to it, whether the ship were moving uniformly or at rest, because the motion of the ship, though unnoticed, was also conveyed to the falling object. It is the same with a stone dropped from a tower which falls parallel to the tower despite the motion of the earth during its fall. As Salviati states:

> With respect to the earth, the tower, and ourselves, all of which keep moving with the diurnal motion along with the stone, the diurnal movement is as if it did not exist; it remains insensible, imperceptible, and without any effect whatever. All that remains observable is the motion which we lack, and that is the grazing drop to the base of the tower. (p. 171)

A similar argument applies to cannon balls fired on the earth at the same moment in opposite directions with identical elevation and force. It would seem that while the cannon balls were in flight the distance of the ball fired in the opposite direction to the earth's motion would be greater because its duration would include the sum of the two motions, the measured westward motion of the cannon ball plus the eastward motion of the earth carrying the cannon during its flight. In contrast, the one fired in the same direction as the earth's motion would be less because its trajectory would include the measured distance of its motion eastward minus the forward motion of the cannon carried by the earth's motion while the ball was in flight. But this too ignores the fact that the motion of the earth is not independent of the other motions and thus cancels out.

In describing these examples I was struck by how easy it is to derive the opposite conclusions from different assumptions and how reasonable each seems. Thus one can sympathize with Simplicius who found it difficult to accept Salviati's argument. They also illustrate why revolutionary scientific discoveries incite so much opposition because of initially appearing counterintuitive in relation to previously accepted assumptions. And since all of our intuitions reflect a background framework of interpretation involving familiar experiences and common sense beliefs, the initial reaction to revolutionary developments is to reject them, as was true of the heliocentric theory, the first geological discoveries, Darwin's conception of evolution, Max Planck's quantum theory, and Werner Heisenberg's uncertainty principles.

This is especially true for religionists who are brought up to believe in infallible revelations and to accept things on Church authority or unquestioned faith, both of which are antithetic to the scientific attitude. But for those like Kepler and Galileo who believed that scientific inquiry is the most trustworthy method for attaining *empirical knowledge* and who were committed to accepting objective evidence, giving up traditional beliefs when there was sufficient evidence was not as difficult. As Sagredo declares,

> considering that everyone who followed the opinion of Copernicus had at first held the opposite, and was very well informed concerning the arguments of Aristotle and Ptolemy, and that on the other hand none of the followers of Ptolemy and Aristotle had been formerly of the Copernican opinion and had left that to come round to Aristotle's view…I commenced to believe that one who forsakes an opinion which he imbibed with his [mother's] milk and which is supported by multitudes, to take up another that has few followers and is rejected by all the schools and that truly seems to be a gigantic paradox, must of necessity be moved, not to say compelled, by the most effective arguments. (pp. 128–129; brackets added)

"That he represented it to be an argument for the truth that Ptolemaics become Copernicans, but not vice versa" (p. 477, note 103:7), one of the eight propositions seized upon by the commission of the Inquisition appointed by the pope as points offensive to the Church, was taken from the above quotation.

Following the second day's dialogue directed against the opposition to the earth's diurnal rotation on its axis, the third day is addressed to refuting the arguments opposed to the earth's annual revolution around the sun. Because it would displace the Earth from the center of the universe, Simplicius is of course opposed to this second motion. Salviati responds by pointing out that having a center presupposes that the cosmos is a finite sphere, yet no one "has so far proved whether the universe is finite and has a shape, or whether it is infinite and unbounded." (p. 319) But because "Giordano Bruno's conviction and execution had depended largely upon his having espoused the view that the universe was infinite [along with doubting the truth of the Trinity and the sacrament of the Eucharist]" (p. 493, note 319; brackets added), the discussion is not pursued.

But assuming that the cosmos is spherical and thus has a center, Salviati asks whether this corresponds to the center of the earth or to the celestial orbs, with Simpliciius responding by asking why the two could not coincide? This allows Salviati to return to his argument that, given Kepler's three astronomical laws and Galileo's telescopic observations, the simplest and most harmonious astronomical system is achieved by assigning the central position to the Sun, not the Earth. For instance, as seen from the Earth Mars' distance from the Earth varies so radically that its orbit could not possibly be circular as required by the Aristotelian system of concentric spheres. As Salviati adds, there is additional consilient evidence for believing that the sun is in the center.

> This is reasoned out from finding the three outer planets—Mars, Jupiter, and Saturn—always quite closer to the earth when they are in opposition to the Sun, and very distant when they are in conjunction with it. This approach and recession is of such moment that Mars when close looks sixty times as large as when it is most distant. Next, it is certain that Venus and Mercury must revolve around the sun, because of their never moving far away from it, and because of their being seen now beyond it and now on this side of it, as Venus's changes of shape conclusively prove. (p. 322)

As if this were not sufficient evidence, Salviati notes that Galileo's telescopic observations indicate that the orbits of Mercury and Venus are below the orbit of the Earth and around the Sun, while those of Mars, Jupiter, and Saturn are above the earth's orbit. Given this evidence for the Copernican system, Sagredo asks why, "[i]f this very ancient arrangement of the Pythagoreans [Philolaus and Heraclides of Pontus] is so well accommodated to the appearances…it has found so few followers in the course of centuries" (p. 327; brackets added), to which Salviati gives the reply:

> No, Sagredo, my surprise is very different from yours. You wonder that there are so few followers of the Pythagorean opinion, whereas I am astonished that there have been any up to this day who have embraced and followed it. Nor can I ever sufficiently admire the outstanding acumen of those who have taken hold of this opinion and accepted it as true; they have through sheer force of intellect done such violence to their own senses as to prefer what reason told them over that which sensible experience plainly showed them to the contrary…that Aristarchus and Copernicus were able to make reason so conquer sense that, in defiance of the latter, the former became mistress of their belief. (pp. 327–328)

Simplicius still not being convinced, Salviati considers other objections offered by the scholastics, such as why it is just in the Copernican system that only the Moon revolves around the Earth with all the other planets revolving around the Sun? To begin with, he cites Galileo's telescopic discovery of the four Jovian moons, that he named the Medicean stars, which removes this "apparent anomaly of [only] the earth and the moon moving conjointly." (p. 340; brackets added) He then addresses two other objections, the apparent retrograde motion of the five planets and the stellar absence of parallax. As to the first, he refers to a diagram drawn by Galileo depicting how the annual revolution of the earth around the sun produces the illusion of the loop-like retrogressions of the planets as seen from the earth's revolution, thus helping Sagredo understand how "these stoppings, retrograde motions, and advances, which have always seemed to me to be highly improbable," (p. 342) are rejected by the Copernican system as illusory.

The objection from parallax is that if the earth revolved around the sun, then during its orbit one should observe a displacement of the fixed stars in the opposite direction, just as trees seen from a moving carriage appear to recede laterally. The usual explanation that this is due to their great distance is rejected by the Aristotelians on the grounds that "in order for a fixed star to look as large as it does, it would actually have to be so immense in bulk as to exceed the earth's orbit—a thing…entirely unbelievable." (p. 372; brackets added) There being no evidential rebuttal, Salviati offers the common sense reply that not enough is known about how the stars transmit their light from such a great distance to draw any definite conclusion.

He then ends with a summary of all the convincing evidence showing the superiority of the heliocentric position even though Pope Urban VIII decreed that he was free to write about the two systems of the world *provided he treated them impartially.*

> See, then, how two simple noncontradictory motions assigned to the earth, performed in periods well suited to their sizes, and… conducted from west to east as in the case of

all movable world bodies supply adequate causes for all the visible phenomena. These phenomena can be reconciled with a fixed earth *only by renouncing all the symmetry that is seen among the speeds and sizes of moving bodies, and attributing an inconceivable velocity to an enormous sphere beyond all the others*, while *lesser spheres move very slowly*. Besides, one must make the motion of the former contrary to that of the latter, and to increase the *improbability*, must have the highest sphere transport all the lower ones opposite to their own inclination. I leave to your judgment which has the more likelihood in it. (p. 396; italics added)

Thus despite Urban's earlier assertion "of there being no fear anyone would prove it true," Galileo clearly affirms the "improbability" of geocentrism in comparison to heliocentrism. While it could be argued that he treated the evidence of the two systems impartially, as Urban insisted, in the sense of being fair and unprejudiced, his obvious affirmation of the evidence for the Copernican system, as against the Aristotelian, does not suggest a commensurate or equivalent evaluation, as I think was the intended meaning of the pope. This will prove to be the major factor leading to his condemnation at the subsequent trial.

It is ironic that despite the sagacity of the previous arguments the fourth and final dialogue containing what Galileo then believed to be the most convincing evidence for the heliocentric system, that the two motions of the earth explained the tides, was false and disclaimed by all other astronomers. He had been so convinced by it that he had wanted to insert 'tides' in the title of the book, but was prevented from doing so by Urban. The conjunction of the ebb and flow of the tides with the orbital motion of the moon was well known. As previously recounted, Kepler in his *Astronomia Nova* had attributed the tides to the mutual gravitational attraction of the moon and the earth, but Galileo's disdain for mysterious forces acting at a distance, reminiscent of occult powers, led him to reject this explanation in favor of a mechanical one due to the two contrasting motions of the earth, illustrating that even the most gifted scientists can succumb to false explanations.

As recounted preiously, drawing an analogy with the pitching of ships in rough seas causing any water in the bilge to flow back and forth, he proposed that the shift in the ocean's basins produced by the contrasting motions of the earth caused the tides. Sagredo declares that having

> read and listened to the great follies which many people have put forth as causes for these events, I have arrived at two conclusions…that if the terrestrial globe were immovable, the ebb and flow of the oceans could not occur naturally; and that when we confer upon the globe the movements just assigned to it, the seas are necessarily subjected to an ebb and flow agreeing in all respects with what is to be observed in them. (p. 417)

In conclusion, Sagredo expresses near the end of the *Dialogue* his confidence in the superiority of the Copernican system, claiming three kinds of evidence have been shown to be very convincing.

> In the conversations of these four days we have, then, strong evidences in favor of the Copernican system, among which three have been shown to be very convincing—those taken from the stoppings and retrograde motions of the planets, and their approaches toward and recessions from the earth; second, from the revolution of the sun upon itself, and from what is to be observed in the sunspots; and third, from the ebbing and flowing of the ocean tides. (p. 462)

Perhaps realizing that he had been too strong in defending the Copernican system—though given his belief in the soundness of the evidence and his integrity and striving for the truth it is hard to see how he sincerely could have done otherwise—he apparently tried to soften the effect by having Simplicius, the rebutted scholastic, express the admonition of Urban at the very end of the *Dialogue*.

> As to the discourses we have held, and especially this last one concerning the reasons for the ebbing and flowing of the ocean, I am really not entirely convinced; but from such feeble ideas of the matter as I have formed, I admit that your thoughts seem to me more ingenious than many others I have heard. I do not therefore consider them true and conclusive; indeed, keeping always before my mind's eye a most solid doctrine that I once heard from a most eminent and learned person [Pope Urban VIII] and before which one must fall silent, I know that if asked whether God in His infinite power and wisdom could have conferred upon the watery element its observed reciprocating motion using some other means than moving its containing vessels, both of you would reply that He could have, and that He would have known how to do this in many ways that are unthinkable to our our minds. From this I forthwith conclude that…it would be excessive boldness for anyone to limit and restrict the Divine power and wisdom to some particular fancy of his own. (p. 464; brackets added)

If he thought this would assuage Urban's vehement negative reaction he was tragically mistaken as events will show, especially as his enemies pointed out to the pope that his disclaimer was said by Simplicius, the rebuked scholastic, rather than by Sagredo, the arbitrator in the dispute, although it is arguable that it was more appropiately said by Simplicius whose views were aligned with the pope's. Furthermore, although including the three previous arguments as also those he was not "entirely convinced" of, his disclaimer explicitly refers to his argument about the tides, recalling Urban's doctrine that "God in his infinite power could have conferred upon the watery element its observed reciprocating motion using some other means than its containing vessel." While expressing reservations about this argument, he left little doubt that he had found the three previous arguments "very convincing."

A concluding statement by Salviati raises a further question as to Galileo's actual conviction at this time regarding the veracity and future progress of scientific knowledge which he had proclaimed throughout the *Dialogue*. Was he sincere in this final avowal of skepticism or was it one final effort to placate Urban despite his previous assertions—or had he not actually resolved the dilemma even though in his previous writings he seemed to have done so? When so engrossed in writing a book, especially one like his comparing two very controversial cosmological systems with its dire religious consequences, it is easy to be unduly swayed at the time by one's own arguments. Referring to the previously quoted passage by Urban to not letting one restrict the "Divine Power and Wisdom" by one's own views, Salviati states:

> An admirable and angelic doctrine, and well in accord with another one, also Divine, which, while it grants to us the right to argue about the constitution of the universe (perhaps in order that the working of the human mind shall not be curtailed or made lazy) adds that *we cannot discover* the work of His hands. Let us, then, exercise these activities permitted to us and ordained by God, that we may recognize and thereby so much the more admire His greatness, however much less fit we may find ourselves to penetrate the profound depths of His infinite wisdom. (p. 464; italics added)

Here the contradiction is more apparent in that while we "are ordained by God" to pursue these questions "to admire His greatness," we shall never find the answers.

The difficulties of the book being published having been overcome after two years of negotiations, the contrasting reactions were glaring. When it appeared in February 1632 it was fervently praised by his friends, Castelli writing that "I still have it by me, having read it from cover to cover to my infinite amazement and delight; and I read parts of it to friends of good taste to their marvel and always more to my delight, more to my amazement, and with always more profit to myself." (Drake, pp. 336–337)

The hostile reaction from the clergy was just as fervent, the Jesuits extremely aroused demanding that the book be prohibited while Urban was furious believing that the assured manner in which Galileo had defended the Copernican system, despite his later expressions of reservations as to its truth, belied Urban's own cautionary statements that it was unlikely that anyone "could prove it true." He was especially incensed when he learned that the commission of the Inquisition had informed Galileo that "he must not hold, defend, or teach in anyway orally or in writing the motion of the earth and stability of the sun, on pain of punishment."

He believed he had been deliberately deceived and betrayed by Galileo's not having informed him of the restraining Edict of 1616 during their friendly meeting in April 1624 when he gave his permission to write his book.

As a result the Florentine Inquisitor, who had approved the book for publication, sent the original manuscript the following September to Rome for examination, ordered Galileo to appear before the Inquisition in Rome the following November, and proscribed all sales of the book. Being ill at the time, he did not appear in Rome until 13 February of the following year. Despite the hostile reaction of Urban and the Jesuits, Galileo apparently believed that the affidavit of assurance that Bellarmine had given him, which his accusers were unaware of, would absolve him. Yet not withstanding my enormous admiration for Galileo which should be obvious and loathing of the repression of the Catholic Church and the cruelty of the Inquisition, it is difficult to see how Galileo could have exonerated himself. Recall that on 26 of February 1616, as recounted by Drake,

> Bellarmine told Galileo of the official finding against the motion of the earth and stability of the sun and told him he must not hold or defend them any longer and that immediately and without any intervening relevant event...Bellarmine being still present, the commissary of the Inquisition admonished Galileo in the name of the pope that he must not hold, defend, or teach in any way, orally or in writing, the said propositions on pain of imprsonment. Galileo agreed. (p. 253)

In all fairness how could he have claimed that he had not defended "the motion of the earth and stability of the sun" in his *Dialogue* and in correspondence to friends? Sagredo had forthrightly declared that he was "very convinced" by three different kinds of evidence supporting the heliocentric theory. Even at the beginning of the *Dialogue* Galileo had frankly stated that "*I have taken the Copernican side in the discourse, proceeding as with a pure mathematical hypothesis and striving by every artifice to represent it as superior to supposing the earth motionless—not, indeed, absolutely, but as against arguments of some professed Peripatetics.*" (pp. 5-6; italics in original) While he indicated that his approach was mathematical and not demonstrated *absolutely*, in conformity with Urban's instruction's, he obviously did not treat the two systems impartially as he had been directed.

Nor did the meeting with Urban VIII in which he received permission to write a book on the two systems, provided he treated them equally and not go beyond the astronomical and mathematical arguments, exonerate him since he had not informed Urban of the admonition of the commission of the Inquisition nor of the Edict of 1616. The latter was his fatal omission for this was the basis of the accusations against him, along with his obvious preferential arguments for heliocentrism in the *Dialogue*.

Having been housed in the Tuscan Embassy since his arrival in Rome, on 12 April he was transferred to the Offices of the Inquisition where he was treated cordially and placed in a comfortable apartment with his attendant. The interrogation began the next day based on the Edict of 1616 and Bellarmine's affidavit of assurance. As recounted by Drake, in the affidavit Bellarmine had absolved Galileo of having "abjured…any doctrine or of having been given any salutary penance," but

> was only told of the declaration made by his Holiness [Pope Paul V] and published by the Congregation of the Index, that [to hold that] the earth moves around the sun and that the sun stands still in the center of the universe without motion from east to west is contrary to Sacred Scripture and therefore may not be defended or held. (p. 348; brackets added)

When Galileo insisted that he had not contravened the precept in any way, Vincenzo Maculano, the Commissary General of the Inquisition, realizing that this was not true, proposed "'that the Holy Congregation grant me power to deal extrajudicially with Galileo to the end of convincing him of his error and bringing him to the point of confessing it when he understood.'" (p. 349) Then in a private meeting Maculano did persuad him that his *Dialogue* contravened the Edict and even Bellarmine's affidavit. Finally conceding, Galileo asked for time to reconsider his book and write his confession which he submitted at the end of April. As Drake retells this,

> having reread it he realized that in many places a reader ignorant of his intention might think the arguments earned the day for the position *he meant to confute*, especially the arguments from the sunspots and from the tides…. He had not meant any disobedience but confessed vain ambition, ignorance, and inadvertence. (p. 350; italics added)

If Drake's account is accurate, one cannot help but wonder what his mental state was if he could claim now to have "*meant to confute*…the arguments from the sunspots and from the tides…." I do not believe he was dissembling or lying, but that the mental torment must have put him in a state of doubt or of self-deception. Or perhaps he had undergone some kind of mental conversion in which he now sincerely disowned all that he had previously defended, such as the veracity of empirical evidence over ancient scripture, rational arguments over faith, and the credence of natural science over religious authority in *questions of empirical fact*. For example, in his earlier Letter to Castelli on December 1613 he had written, again as reported by Drake, that so little is actually said in Scripture about astronomy that not even the planets are named and

> "as we are unable to assert with certainty that all interpreters speak with divine inspiration, I should think it would be prudent if no one were permitted to oblige Scripture…

to sustain as true some physical conclusion of which sense and demonstrative and necessary reasons may show the contrary. And who wants to set bounds to the human mind." (p. 226)

That he underwent a radical change of belief is further indicated in a letter he wrote to Francesco Rinuccini shortly before his death repudiating all that he had previously upheld and fought for.

> "The falsity of the Copernican system must not on any account be doubted, especially by us Catholics, who have the irrefragable authority of...scriptture interpreted by the greatest masters in theology, whose agreement renders us certain of the stability of the earth and mobility of the sun.... The conjectures of Copernicus and his followers offered to the contrary are all removed by that most sound argument taken from the omnipotence of God. He being able to do in many, or rather infinite ways, that which to our view and observation seem to be done in one particular way...." (p. 417)

Thus he renounced his earlier argument that neither scripture itself nor the attestations of divinely inspired interpreters should be accorded greater authority in *scientific questions* than sensory observation and rational demonstrations and in the end succumbed to Urban's argument.

Despite his signed confession, on 16 June, according to Drake, Urban "ordered Galileo's examination on intention, to be followed (if he sustained this) by his imprisonment for an indefinite term at the pleasure of the Holy Office, confiscation of the *Dialogue*, and mandatory public reading of the sentence to professors of mathematics throughout Italy and elsewhere." (p. 351) Examined on 21 June as to his intention, also related by Drake, "'he stated that until the 1616 decree he had considered the two world systems to be freely debatable, but [still insisted] that thereafter he had adhered to the fixed earth and movable sun; in his book he had considered no argument as conclusive and the decision of sublime authority as binding." (p. 351; brackets added)

While it is true that he did not consider his arguments as conclusive, he certainly had not held "the decision of sublime authority as binding." Asked on pain of torture (which was never used) if he spoke the truth, he replied "'that I am here to obey, and have not held this opinion after the determination made, as I said.'" (p. 351) The next day "the sentence of life imprisonment was read to Galileo at a formal ceremony in the presence of the cardinals of the Inquisition and witnesses, after which he had to abjure on his knees before them." (p. 351) Although not as horrific as Giordano Bruno's being burned at the stake for his so-called heretical beliefs, the sorry ordeal of Galileo's trial has justly been called "the disgrace of the century."

But not all the clerics acted disgracefully; Cardinal Barberini, the pope's nephew, who stood by Galileo throughout the trial shielding him from more severe treatment, was one of three cardinals that refused to sign the sentencing document and who intervened to have his imprisonment transferred back to the Tuscan embassy. Then receiving an invitation from Archbishop Ascanio Piccolomini of Siena who was an avid admirer of Galileo, the pope allowed him to accept the invitation and go to Siena in his custody. Initially distraught over the sentencing, Galileo gradually recovered his physical and mental health thanks to the warm reception of the Archbishop and even began writing his second great treatise, *Dialogues Concerning Two New Sciences*. Again owing to the urging of Cardinal Barberini with the support of Ambassador Niccolini, the pope on 1 December "recommended to the Holy Office that Galileo be permitted to return to his villa at Arcetri in the hills some distance from Florence, provided he receive few visitors and refrain from teaching." (p. 356) This too proved to be star-crossed because though it brought him close to the convent of his beloved daughter, Sister Maria Celeste, this precious solace was taken away when she died suddenly the following year.

It was at Arcetri that he completed the *Dialogues Concerning Two New Sciences*[40] in which the three discussants are the same as in his previous dialogue. Though largely a recapitulation of earlier research, he described it as "'superior to most everything else of mine hitherto published'" that "'contain results which set forth a very new science dealing with a very ancient subject.

> There is, in nature, perhaps nothing older than motion, concerning which the books written by philosophers are neither few nor small; nevertheless I have discovered by experiment some properties…which are worth knowing and which have not hitherto been either observed or demonstrated. Some superficial observations have been made, as, for instance, that the free motion…of a heavy falling body is continuously accelerated; but to just what extent this acceleration occurs has not yet been announced; for so far as I know, no one has yet pointed out that the distances traversed, during equal intervals of time, by a body falling from rest, stand to one another in the same ratio as the odd numbers beginning with unity. (p. 147)

Thus he reintroduced his earlier discovery in 1604 of the odd number law describing the acceleration of free falling objects which, unknown to him apparently, had been previously formulated by Nicole Oreme in the 14th century.

Referring next to his continuing investigation of projectile motion he now presents a much more promising conception of the future of science, declaring that his investigations have opened up a whole new field of inquiry.

> It has been observed that missiles and projectiles describe a curved path of some sort; however no one has pointed out the fact that this path is a parabola. But this and other facts, not few in number or less worth knowing, I have succeeded in proving; and what

I consider more important, there have been opened up to this vast and most excellent science, of which my work is merely the beginning, ways and means by which other minds more acute than mine will explore its remote corners. (pp. 147–148)

Rejecting the Aristotelian explanation of projectile motion as due to a continuous and contiguous motive cause, Galileo apparently adopted Buridan's fourteenth century explanation of motion as caused by an impressed force or *impetus* producing what Galileo called *velocitas* or momentum, both of which are quantifiable as the ratios between two magnitudes, space and time. However, he realized that before attempting a causal explanation it was necessary to describe the parameters needed for a precise quantitative description of the motion itself, the exact approach of modern science though the reverse of the Aristotelians who were primarily concerned with first seeking the cause. Again setting an example for the new scientific methodology, he proceeds in his discussion of motion from the simpler to the more complex, dividing it into three parts: the first dealing with uniform motion, the second with accelerated motion, and the third with Aristotle's so-called violent motion.

He even predates the formal mode of presentation of another mind "more acute than mine" (the phrase in the previous quotation), Newton's, in dividing the analysis of uniform motion into "Definitions, Axioms, Theorems, and Propositions." For example, he defines uniform motion as "one in which the distances traversed by the moving particle during any equal intervals of time, are themselves equal" (p. 148), which is followed by four axioms and six theorems with propositions that include linear diagrams to illustrate his reasoning.

Proceeding to naturally accelerated motion, his discussion again initiates the methodology of modern science in that he rejects a priori definitions for a "definition best fitting natural phenomena…and to make this definition…exhibit the essential features of observed accelerated motions" (p. 154), exemplified by his odd number law that he had discovered earlier.

> It is thus evident by simple computation that a moving body starting from rest and acquiring velocity at a rate proportional to the time, will, during equal intervals of time, traverse distances which are related to each other as the numbers beginning with unity, 1, 3, 5; or, considering the total space traversed, that covered in double time will be quadruple that covered during unit time; in triple time, the space is nine times as great as in unit time. And in general the spaces traversed are in the duplicate ratio of the times, i.e., in the ratio of the squares of the times. (p. 170)

Although Simplicius states that he is convinced that matters are as described, he suggests "that this would be the proper moment to introduce one of those experiments—and there are many of them, I understand—which illustrate in several ways the conclusions reached." (p. 171) Given the precise description of the experiment I find it difficult to believe that the following two contemporaries of Galileo doubted or denied that he had performed it, Mersenne asserting "I doubt whether Galileo actually performed the experiments of incline planes, since he does not speak of them" and that Descartes "'denied' all of Galileo's experiments,'" while even the Galilean scholar Alexander Koyré claimed "in spite of Galileo's assertion, one is tempted to doubt it."[41] Either they had not read him or were too prejudiced to acknowledge what he said.

> A piece of wooden molding or scantling, about 12 cubits long, half a cubit wide, and three finger-breaths thick, was taken; on its edge was cut a channel a little more than one finger in breadth; having made this groove very straight, smooth, and polished, and having lined it with parchment, also as smooth and polished as possible, we rolled along it a hard, smooth, and very round bronze ball. Having placed this board in a sloping position, by lifting one end some one or two cubits above the other, we rolled the ball… along the channel, noting…the time required to make the descent. We repeated this experiment more than once in order to measure the time with such an accuracy that the deviation between two observations never exceeded one-tenth of a pulse-beat. (p. 171)

According to Koyré, the skepticism was based on the fact that all of Galileo's experiments…which resulted in measurements, in precise values, were falsified by his contemporaries." (Koyré, p. 107) But this must have been due to their ineptitude, not because Galileo had not performed the experiments nor that they were not exact. Even Koyré, a few lines later, states that "in spite of this it is Galileo who is in the right" (p. 107), as his description of how he performed such precise measurements verifies.

> Having performed this operation and having assured ourselves of its reliability, we now rolled the ball only one-quarter the length of the channel and having measured the time of its descent, we found it precisely one-half of the former. Next we tried other distances, comparing the time for the whole length with that for the half, or with that for two-thirds, or three-fourths, or indeed for any fraction; in such experiments, repeated a full hundred times, we always found that the spaces traversed were to each other as the squares of the times, and this was true for all inclinations of the plane, i.e., of the channel, along which we rolled the ball. (pp. 171–172)

Despite the skeptics, his experiment clearly demonstrated what was impossible to measure directly (because the velocity of a free falling body is less than three seconds from a ten story building), that its acceleration is proportional to the squares of the times, not to its weight as the Aristotelians held. Yet a question remained as to whether this was true also of a freely falling object? His last statement answers this question since free fall is merely the final extrapolation of successively raising the incline plane to a vertical angle. To avoid any further doubts he describes how he measured the times by measuring the ratios of the water accumulated during the decent, a liquid hourglass procedure.

Extrapolating further, he declares that in a vacuum objects would fall at the same velocity despite differences in their weights, not instantaneously as the Aristotelians claimed, having arrived at this conclusion earlier by experiments in hydrostatics.

> Because if we find...that the variation of speed among bodies of different specific gravities [weights] is less and less according as the medium becomes more and more yielding, and if finally in a medium of extreme tenuity, though not a perfect vacuum, we find that, in spite of great diversity of specific gravity [peso], the difference in speed is...almost inappreciable, then we are justified in believing it highly probable that in a vacuum all bodies would fall with the same speed. (p. 70; brackets added)

This conclusion was supported by another experiment suggested by his earlier observations of the synchronous swings of the lanterns in the Cathedral of Pisa. Measuring and comparing the lengths and times of the pendular arcs of a suspended ball of cork and a vastly heavier ball of lead from strings of equal length from the same extended position, he found the lengths and times of the swings to be identical. Thus he concluded that "if these same bodies traverse equal arcs in equal times we may rest assured that their speeds are equal." (p. 82) Noting that this is difficult to accept, Sagredo makes the shrewd but sad observation about human nature.

> Not only this but also many other of your views are so far removed from the commonly accepted opinions and doctrines that if you were to publish them you would stir up a large number of antagonists; for human nature is such that men do not look with favor upon discoveries—either of truth or of falsity—in their own field, when made by others than themselves. (p. 80)

Imagining the perpetual motion of entities propelled along frictionless horizontal surfaces devoid of air resistance, he correctly predicted the conservation of momentum, but not of inertial motion, believing that the pull of gravity would cause the object to fall in a curved trajectory. He also did not distinguish between an object's mass and weight, the latter produced by gravity, as is now familiar from weightless maneuvers in gravity free outer space. Thus an object in inertial motion freed of gravity and other forces will continue to move uniformly in a straight line due to its inherent mass and motion.

As Galileo's example of a frictionless, horizontal, uniform motion subject to the earth's gravity indicates, it did not conform to these specifications.

> Imagine any particle projected along a horizontal plane without friction; then we know, from what has been more fully explained in the preceedding pages, that this

particle will move along this same plane with a motion which is uniform and perpetual, provided the plane has no limits. But if the plane is limited and elevated, then the moving particle, which we imagine to be a heavy one, will on passing over the edge of the plane acquire, in addition to its previous uniform and perpetual motion, a downward propensity due to its own weight; so that the resulting motion which I call projection…is compounded of one which is uniform and horizontal and of another which is vertical and naturally accelerated. (p. 234)

But while not exactly depicting inertial motion, it did enable Galileo to identify projectiles as compounded of the two motions just described that "produce the path of a projectile, which is a parabola." (p. 248) From this he was able to deduce that "[w]hen the motion of a body is the resultant of two uniform motions, one horizontal, the other perpendicular, the square of the resultant momentum is equal to the sum of the squares of the two component momenta." (pp. 246–247) This enabled him to make additional calculations and predictions, such as the angle of elevation that would give the longest trajectory of a cannon ball.

Thus concludes the discussion of one of the seminal works of early modern science because its experimental investigations of various kinds of motion disclosing the mathematical ratios, a crucial feature of the physical and astronomical sciences, prefigures the methods and explanations of later scientists. Though not containing any material proscribed by the Catholic Church, Galileo encountered considerable resistance when he attempted publishing in Florence or Venice, the Venetian Inquisitor pointing out "that there was an express order prohibiting the printing or reprinting of any work of Galileo, either in Venice or any other place, *nullo excepto*." (p. xi)

Despite the pope's two previous benevolent permissions of relocation, this shows the continuing severity of the Holy Office toward him. When he requested to go to Florence for treatment of a painful hernia because no doctors were available in Arcetri, "the request was not only refused, but the Florentine Inquisitor was instructed to tell Galileo that any more such petitions from him would result in imprisonment." (Drake, p. 360) Yet after many inquiries he succeeded in having the book published in 1638 by the founder of one of the great emerging printer-designers in Europe, the distinguished publishing house of Louis Elzevir in Amsterdam that still exists.

Although blind and exceedingly weary, after the publication he lived four more years until his death on 9 January 1642 in Arcetri, less than two months before his seventy-eighth birthday. After his interment in the majestic church of Santa Croce in Florence the Grand Duke sought to erect a grand tomb across from that of Michelangelo, but was prevented by the Catholic Church which "forbade any honors to a man who had died under vehement suspicion of heresy." (Drake,

p. 436) Yet despite the Church's prohibition at the time, there now exists in the church of Santa Croce a monument to Galileo just as splendid as the one honoring Michelangelo.

Although this discussion of the contribution of Galileo to the emergence of modern science has been relatively brief, I hope it has shown how Galileo nearly single-handedly dismantled Aristotle's enduring cosmology, along with his system of inquiry and explanation, replacing them with an explicit new methodology and theoretical framework founded on his remarkable telescopic discoveries and experimentally proven laws of motion. As Clavelin's summary of his achievements state:

> The reason… no scientific problem was ever the same again as it had been before Galileo tackled it lay largely in his redefinition of scientific intelligibility and in the means by which he achieved it: only a new explanatory ideal and…unprecedented skill in combining reason with observation could have changed natural philosophy in so radical a way. No wonder then, that, as we read his works, we are struck above all by the remarkable way in which he impressed the features of classical science upon a 2000-year-old picture of scientific rationality. (p. 383)

Add to this his courageous attempt to free scientific inquiry from the pernicious authority of the Catholic Church (which persists to this day) and one can begin to appreciate his extraordinary achievements.

CHAPTER SEVEN

The Divergent Philosophies of Bacon and Descartes

The two philosophers considered next, Francis Bacon and René Descartes, are not discussed because they made any actual contributions to the emerging methodology and conceptual framework of modern science, but because of their critical appraisal of the still generally inchoate state of scientific inquiry and call for a clean sweep. Both were highly regarded and extremely influential at the time for repudiating the traditional respect for antiquity and advocating a new beginning, though unlike their scientific contemporaries they did not make any genuine contribution to the emerging scientific endeavors. I shall argue that had their own prescriptions for the reconstruction of natural philosophy been followed it would have obstructed, rather than advanced, the creation of modern science.

Due to the discordant nature of scientific inquiry in the transitional period of the 16th and early 17th centuries, owing to the contentious claims by the Aristotelians, Platonists, Neoplatonists, scholastics, alchemists, and astrologists, there was a general awareness of the need for a more effective method of inquiry and means of explanation. But it was Bacon's writings especially that were extensively read and influenial due to their ardent call for a radical revamping of the current disarray of methodologies, along with his astute analysis of the common weaknesses of human thought, that account for his tremendous influence.

Yet renown as a prophet for his expectations for the future of science if his visionary project was fulfilled, compared to the actual founders of modern science, such as Copernicus, Gilbert, Kepler, Galileo, and Harvey, he was incapable of determining what actual changes in methodology were necessary and *how they should be applied to achieve this*. His criticisms of the state of knowledge accompanied by wise aphorisms were sagacious, but his indifference to the discoveries of those contemporaries who were actually advancing scientific inquiry prevented him from discerning and benefiting from the reasons for their success.

These advances included the construction of a new theoretical framework for describing the solar system; the discovery of laws and causal explanations of astronomical and physical motions; the use of experimentation to make new discoveries and test hypotheses; the replacement of Aristotle's logical deduction by mathematics as the correct formalism of science; and the realization that the simplicity and harmony of a theoretical system can be more accurate criteria of physical reality than ordinary sensory observations. Though heralding a new era in science, Bacon rejected Copernicus's heliocentric system, disparaged Gilbert's magnetic discoveries, ignored Kepler's remarkable formulation of the first astronomical laws and clockwork system of the universe, and dismissed Galileo's experiments and mathematical formulation of the law of free fall as unconfirmed. As we are said to learn from past mistakes as well as successes, the reason for discussing Bacon and Descartes is to show why, as gifted as were, they failed to recognize the essential features of scientific inquiry as it was currently being applied and thus were unable to make any real contributions to the current scientific revolution, in contrast to their usual reputations.

Bacon's life's achievements were diverse and outstanding, including his service to the English sovereigns, along with his extensive and extremely influential writings. Born in 1561 into an illustrious and learned family, his father Sir Nicolas Bacon served as "Lord Keeper of the Great Seal and Lord Chancellor, as well as a principle jurist whose counsels Elizabeth heard with respect."[42] His mother, Ann, was the daughter of King Edward VI's tutor and considered "an erudite and upright person, of an intense piety that could not but have had some effect on her sensitive son." (p. 1) He entered Trinity College, Cambridge at age twelve and Gray's Inn four years later to study law. He then sought a career in Parliament in 1584 that would lead to continuous advancement and an illustrious career, first under Queen Elizabeth I and then under King James I who bestowed on him many honors and titles, such as Baron Verulam and then Viscount of St. Albans.

Sadly, this distinguished career of public service came to an end in 1621 when the House of Lords found him guilty of corruption, though this did not seem to have detracted from his public esteem. As Ben Jonson wrote: "I have, and do reverence him for the greatness that was only proper to himself, in that he seemed to me ever, by his work, one of the greatest men, and most worthy of admiration, that had been in many ages. (p. 3)

Sidney Warhaft, the editor who wrote the Introduction to the *Works* from which I am quoting, "is even more emphatic in his praise."

> Whatever Bacon's personal attributes were, there can be no doubt about the respectability of his public service. His dedication to his sovereign, his attempts to revise and codify the laws of his country, his opposition to excessive taxation, his encouragement of a generous colonial policy, and his efforts to weld Great Britain into a unity, all of these are matters of record whose worth posterity has vindicated. (p. 3)

Despite the pain of his conviction and dismissal from parliament, this allowed him to turn to his primary endeavor, the reflections that produced his numerous philosophical writings including his *Essays, The Proficience and Advancement of Learning, The Wisdom of the Ancients, The Great Instauration, The New Organon, Of the Dignity and Advancement of Learning*, and the *New Atlantis*. What he proposes is a revolution of thought, a turning from the reverence of the past to the promise of the future built on a reconstruction of knowledge involving a new methodology driven by faith in progress and hope for the future—a foreshadowing of the Enlightenment that was considerably influenced by his writings.

Yet he (with Descartes) is one of the clearest examples of the difficulty for most people of extricating their thinking from the currently accepted modes of thought and attempting to replace them by others more credible. But the awareness of faults is not the same as discerning the means of correcting them! Bacon is brilliant as regards the former, which accounts for his extraordinary international reputation, but deficient in the latter. Warhaft claims that "Bacon was nothing less than the spokesman for and leader of the scientific view of modern times." (p. 8) Granted that he could be regarded as "the spokesman for" the need of an entirely new science, yet he certainly could not be considered the "leader" because he had no influence on actually creating the methodology and conceptual framework of modern science. He is like a gifted cheerleader who cannot perform on the playing field.

If anyone were selected for that title it would be either Galileo because of his telescopic discoveries, brilliant defense of the heliocentric system, experimental proof of the law of free fall, destruction of the Aristotelian system, and tragic attempt to free scientific inquiry from the grip of the Catholic Church. The other choice would be Newton because of the tremendous influence of his research and writings, surpassing anyone in the 16th or 17th centuries, that largely served as the basis of scientific inquiry throughout the following centuries until modified by Einstein's theory of relativity and Planck's introduction of quantum mechanics in the early 20th century.

Turning to Bacon's influence and writings, it was his criticisms of the current state of knowledge and partial analyses of their causes, such as the common failures of human reasoning described in the "Idols," the falsities of Aristotle's system, the sterility of scholastic thought, along with the fallacies of alchemy and astrology that are impressive, not his own prescriptions for correcting them. Except for his call for experimentation and use of mathematics, the need for which he never explained, his proposals for reconstructing knowledge in *The Great Instauration* and *The New Organon* do not indicate an understanding of what was needed to bring about the necessary changes or of the current scientific developments that were achieving this. He decries the huge gap between the scientific attainments of the ancient Greeks and the confused state of knowledge at the time, but failed to appreciate or acknowledge how the efforts of Copernicus, Gilbert, Kepler, Galileo, and Harvey were overcoming this.

To begin with his criticism of the current state of knowledge presented in the *Preface* to *The Great Instauration* published in 1620, he declares:

> *That the state of knowledge is not prosperous nor greatly advancing; and that a way must be opened for the human understanding entirely different from any hitherto known, and other helps provided, in order that the mind may exercise over the nature of things the authority which properly belongs to it.*" (p. 302; italics in the original)

Indicative of the dominant influence of Christianity on his thought, he believed that part of the justification for his reconstruction of knowledge would be the return of mankind to Adam's state when he had "dominion over creation" before God's curse and his fall: "For man by the fall fell at the same time from his state of innocency and from his dominion over creation. Both of these losses however can even in this life be in some part repaired, the former by religion and faith, the latter by arts and sciences.'" (p. 21) But what did Bacon propose as the way to redeem this lost "dominion over creation"?

This is described in the *Preface* to *The New Organon*, the second part of *The Great Instauration*, that he wrote to replace the *Organon* of Aristotle containing his logical treatises.

> I propose to establish progressive stages of certainty. The evidence of sense helped and guarded by a certain process of correction I retain. But *the mental operation* which follows the act of sense I for the most part reject; and instead of it *I open and lay out a new and certain path for the the mind to proceed in*, starting directly from the simple sensous perception the necessity of this was felt no doubt by those who attributed so much importance to Logic, showing thereby that they were in search of helps for the understanding, and had no confidence in the native, and spontaneous, process the mind. (p. 327; italics added)

Here he asserts that it is not the senses if corrected that are the obstacles to attaining knowledge, but the acts of the mind, such as logical inferences and deductions. (p. 327; italics added)

As he goes on to say:

> "There remains but one course for the recovery of a sound and healthy condition; namely that the entire work of the understanding be commenced afresh, and the mind itself be from the very outset not left to take its own course, but guided at every step, and the business be done as if by machinery." (p. 327)

Recognizing the difficulty of the task, he (as Galileo) modestly claims that this will be accomplished by others over an extended time—although in other places he states that this will be attained in the near future if one follows his directives.

> The completion however…is a thing both above my strength and beyond my hopes. I have made a beginning of the work—a beginning, as I hope, not unimportant; the fortune of the human race will give the issue—such an issue, it may be, as in the present condition of things and men's minds cannot easily be conceived and imagined. (p. 323)

The crucial question is whether Bacon did prescribed the means "to open a new way for the understanding" that he recognized was needed and so ardently sought? (p. 328)

Book I of *The New Organon*, "Concerning the Interpretation of Nature and the Kingdom of Man," presents his Aphorisms for achieving this redirection of the mind, including his famous "Idols" describing the "adventitious or innate" dispositions of humans that are a hindrance to attaining true knowledge, such as imposing more order on nature than there is. These Idols "which beset men's mind" are dived into four classes. (Cf. pp. 335–338) The first are the Idols of the Tribe which "have their foundation in human nature itself and in the tribe or race of men." Though not cited by Bacon, these would include all those perceptual and cognitive limitations and distortions due to the natural endowments of our species that gave rise to such false theories as geocentrism, along with susceptibilities to believing in myths, superstitions, miracles, and revelations. The kinds of beliefs that the creators of modern science had to reject and replace.

The Second are the Idols of the Cave that "are the idols of the individual man" rather than the species. These would include personal differences in intelligence or rationality, emotional needs and stability, family and cultural background, inherent talents predisposing one to studies in literature or science, music or mathematics, philosophy or physics, anthropology or astronomy, and the cognitive differences in the appeal and credibility of a supernatural versus a naturalistic worldview that has produced so much controversy in the past and in the present.

Third are the Idols of the Market Place "formed by the intercourse and association of men with each other" facilitated by the deceptive use of language in conveying misinformation, misrepresentation, or guile, so that the "ill and unfit choice of words wonderfully obstructs the understanding." This would be illustrated in the propaganda used by tyrants, autocrats, and sycophants to maintain their power or influence. It is also the reason autocratic governments repress free speech and control the press.

The fourth are the Idols of the Theatre "which have immigrated into men's minds from the various dogmas of philosophies and also from wrong laws of demonstration," as illustrated by all the false philosophies, ideologies, and theories that have misled human thought, such as "the divine right of kings," Nazism, fascism, and communism. Though Bacon refers to "dogmas" that one usually associates with religious doctrines, he does not refer to religion and theology as one of the primary sources of fictitious beliefs. This is because he was a firm believer despite his general critical attitude regarding the contemporary intellectual culture, even accepting the myth of genesis as true (thus inadvertently offering himself as an example of the influence of the Idol of the Tribe on his thought). Not only did he believe the Genesis story of original sin, he accepted the creation account that "God on the first day of creation created light only" before creating the heavenly bodies (p. 350), an especially naïve belief for the time.

As Galileo, he accepts the distinction between divine revelation in contrast to sensory knowledge of nature: "For all knowledge admits of two kinds of information, the one inspired by divine revelation, the other arising from the senses." (p. 411) Yet he is not entirely credulous: "But as for narrations touching the prodigies and miracles of religions, they are either not true or not natural, and therefore impertinent [irrelevant] for the story of nature." (p. 399; brackets added) He approves of Theology, asserting that it "consists either of Sacred History or of Parables, which are a divine poesy, or of Doctrines and Precepts, which are perennial philosophy." (p. 395)

According to Warhaft, he held that "all rational consideration of such problems as deity, the after-life, and the ends of creation is relegated to faith, where 'the more discordant...and incredible the Divine mystery is, the more honour is shown to God in believing it.'" (p. 23) Resembling the doctrine of Tertullian, one of the early church fathers, "that I believe because it is absurd," this is hardly what one would expect from someone who professes a radical reconstruction of knowledge. This reveals an inherent weakness in *his* thinking, in that though he is extremely critical of the current secular knowledge, he seems indifferent to the absurdities of religion, assuming them to be true because they are based on alleged revelation or scripture.

Returning to the question of whether he actually conceived of a method that not only would free mankind from the Idols, but progressively lead to actual knowledge, he states that the foundation of this reconstruction "must be laid in natural history, and that of a new kind and gathered on a new principle." (p. 319) He also asserts that "my history differs from that in use (as my logic does) in many things: in end and office, in mass and composition, in subtlety, in selection also and setting forth, with a view to the operations which are to follow." (p. 319) Despite this glowing description and reference to a "new principle," along with declaring that his "natural history" differs from those in use, one looks in vain for any precise description as to what they actually are, rather than mere embellishments.

There is nothing in Bacon comparable to Gilbert's extensive experimental discoveries of magnetic forces and poles; the remarkable cognitive development of Kepler culminating in his acceptance of elliptical orbits and discovery of the three planetary laws, plus his conception of the cosmos as a mechanized clockwork; nor of Galileo's experimental proof of the law of free fall and precise "redefinition of scientific intelligibility," including his famous statement "that the language of nature is mathematics."

Though Bacon also acknowledged the importance of mathematics in scientific investigations writing that "inquiries into nature have the best result when they begin with physics and end in mathematics" (p. 384), he apparently did not know why this was so criticizing Galileo for representing motion spatially despite the fact that his inclined plane experiments enabled him to discover the laws of gravitational fall and projectile motion based on spatial (and temporal) proportions. Also, Kepler's laws of planetary motion describe the geometrical shapes of the orbits of the planets, along with the ratio of their distances from of the sun and orbital periods due to the sun's influence. Had Bacon read either Kepler or Galileo he would have been aware of the computational advantages of relating motion to spatial magnitudes.

A fuller discussion of his avowed method and axioms does occur in *The Great Instauration* where he contrasts his method with the deductive reasoning of the scholastics based on Aristotle's *Organon*, which he criticizes as mainly suited for disputation, not for seeking new knowledge as he intends in *his Organon*. However, while this *was* largely true of the scholastics it was not true of Aristotle whom he includes in his criticism. Aristotle's method had its failings yet his syllogistic method was not introduced for disputation nor was his inductive inferences based on enumeration as Bacon claims, but to prove more specific empirical facts by logical deduction from sound universal premises derived from inductive evidence.

Nor was Aristotle's method limited to deductive proofs, but included dissections of animals and anatomical studies of mammals and human corpses that were highly praised by Darwin. Thus Bacon's statement that Aristotle "corrupted natural philosophy by his logic: fashioning the world out of categories" [species and genus] (p. 345; brackets added) is a misunderstanding of his contribution to the development of science. These species and genus distinctions did constitute his formal cause and did apply to his biological investigations, but not to his cosmology and natural philosophy. Had Bacon's assessment been true, Aristotle's cosmology would not have had the tremendous influence it did for centuries, despite eventually proving false.

Furthermore, in contrast to Bacon's evaluation Galileo describes Aristotle as "'a man of brilliant intellect'" while Darwin wrote that "'from quotations which I had seen, I had a high notion of Aristotle's merits, but I had not the most remote notion what a wonderful man he was. Linnaeus and Cuvier have been my two gods…but they were more school-boys, compared to old Aristotle'" (these quotations and sources are taken from my book *Seeking in the Truth*, p. 99). Considering that Galileo and Darwin are considered giants of science because of their substantial discoveries, while Bacon did not make a single actual contribution, it is not difficult to decide whose judgments were correct.

The closest Bacon comes to describing his own approach is contrasting it with the deductive method of the scholastics which he describes as "barren of works, remote from practice, and altogether unavailable for the active department of the sciences" (pp. 314–315), ignoring the importance of some of their discoveries. In contrast, he writes that "I on the contrary reject demonstration by syllogism, as acting too confusedly and letting nature slip out of its hands." (p. 314)

> For the end which this science of mine proposes is the invention not of arguments but of arts, not of things in accordance with principles but of principles themselves, not of probable reasons but of designations and directions for works. And as the intention is different, so… is the effect, the effect of the one to overcome an opponent in argument, of the other to command nature in action. (p. 314)

Typically, nothing said specifically describes what the "principles" are that will bring bring about the "arts" that will "command nature in action," just his usual boastful assurance that they will do so. He does not seem to realize that empirical laws are the principles of science.

This is true to of his proposed method of induction that is crucial to his recon-reconstruction of knowledge.

> Now what the sciences stand in need of is a form of induction that shall analyse experience and take it to pieces, and by a due process of exclusion and rejection lead

> to an inevitable conclusion.... Nor is this all. For I also sink the foundations of the sciences deeper and firmer; and I begin the inquiry nearer the source than men have done heretofore, submitting to examination those things which the common logic takes on trust. (pp. 315–316)

But surely the description of his induction as consisting of "analyzing experience" to "take it to pieces" by a process of "exclusion and rejection" must be one of the most imprecise methods ever offered. While scientific inquiry certainly involves the analysis of phenomena, the claim that this should be done by a process of "exclusion and rejection" adds little to the understanding. This "process" consists of the analysis of the observational data and the experimental results, as illustrated previously with examples from the founders of modern science.

He describes the method that he thinks will supplement the senses to bring about his reconstruction of knowledge, experimentation, but without examples.

> To meet these difficulties, I have sought on all sides diligently...to provide helps for the sense—substitutes to...its failures, rectifications to correct its errors; and this I endeavour to accomplish...by experiments. For the subtlety of experiments is far greater than that of the sense itself, even when assisted by exquisite instruments; such experiments, I mean, as are skillfully and artificially devised for the express purpose of determining the point in question. (p. 316–317)

Insisting that experiments are essential to acquiring knowledge, no explanation is given to show why this is so, similar to his usual extravagant claims regarding what he expected to achieve from his new principles. The closest he comes to an exact description of how progress in science will be achieved is his conception of the determinaation of the inner configuration of bodies:

> ...the investigation and discovery of the *latent configuration* in bodies is a new thing.... For as yet we are but lingering in the outer courts of nature, nor are we preparing ourselves a way into her inner chambers. Yet no one can endow a given body with a new nature, or successfully and aptly transmute it into a new body unless he has attained a competent knowledge of the body so to be altered or transformed. Otherwise, he will run into methods which, if not useless, are at any rate difficult and perverse and unsuitable to the nature of the body on which he is operating. It is clear therefore that to this also a way must be opened and laid out. (p. 382)

While stressing the difficulty, this insight is marred by his own "perverse and unsuitable" conception of the "way" he expected it to be carried out.

> For example, we must inquire what amount of spirit there is in every body, what of tangible essence; and of the spirit, whether it be copious and turgid, or meager and

scarce, whether it be fine or coarse, akin to air or to fire, brisk or sluggish, weak or strong, progressive or retrograde, interrupted or continuous.... (p. 383)

One could hardly find a better statement to exemplify his lack of scientific aptitude and ignorance of the current research in science, than this proposal that discovering the "*latent* configuration of bodies" consisted of inquiring "what amount of spirit there is in every body" and its "tangible essence." What did he expect to find by carrying out such a program, especially if we consider the atomic theories of Democritus and Empedocles which were known at the time? Comparing this absurd suggestion with the kinds of investigations being carried out by the previously mentioned scientists ought to have been sufficient to topple Bacon from his pedestal, but it definitely was not.

Not only did he misconceive what would prove in the following centuries to to be the most impressive advances in the history of science, the discovery of the inner structure of matter in terms of the atomic-molecular theory, he also rejected Copernicus' heliocentric hypothesis. In *The New Organon* [43] he initially identifies the problem as to which should be considered real, the two motions attributed to the earth by Copernicus and "much talked about by Philosophers" to explain the astronomical phenomena, or those based on our ordinary observations traditionally believed:

> ...let the nature investigated be the spontaneous motion of rotation, and in particular whether the diurnal motion whereby to our eyes the sun and the stars rise and set, be a real motion of rotation in the heavenly bodies, or a motion apparent in the heavenly bodies, and real in the earth.... Again, let the nature in question be that other motion of rotation so much talked of by philosophers, the...contrary motion to the diurnal, viz., from west to east, which old philosophers attributed to the planets, also to the starry sphere, but Copernicus and his followers to the earth as well. (Aphorisms, Book Two, XXXVI, pp. 194–195)

Bacon then assess whether the observed motions should be considered real or an artifice of calculation: "let us inquire whether any such motions be found in nature, or whether it be not rather a thing invented and supposed for the...convenience of calculation.... For this motion in the heavens is by no means proved to be true and real...." (p. 195)

While this assertion does not indicate his conclusion, his later correct rejection of Galileo's explanation of the tides as due to the two contrasting motions of the earth also includes his *false denial of the two motions* attributed to the earth by Copernicus:

> ...it was upon this inequality of motions...that Galileo built his theory of the flux and reflux of the sea, supposing that the earth revolved faster than the water could follow

and that the water therefore first gathered in a heap and then fell down, as we see it do in a basin of water moved quickly. By this he devised...an assumption which cannot be allowed, viz., *that the earth moves*.... (XLVI; p. 227; italics added).

Since *The New Organon* was published in 1620, eleven years after Galileo began his telescopic observations supporting Copernicus' hypothesis that drew international acclaim, only those astronomers who were blinded by their devotion to Aristotle (recall that Cremonini refused to look through Galileo's telescope because his observations were not mentioned by Aristotle) or to church dogma still adhered to the geocentric position. As Bacon was very critical of Aristotle it must have been his strong religious beliefs, evidenced previously in his assertion that "spirit" constituted the inner nature of objects, that was the basis of his objection to heliocentrism, perhaps owing to the influence of his mother's piety. This also would explain his zealous advocacy of scientific inquiry without knowing what was required to fulfill it.

Despite his scientific ineptitude, it was mainly Bacon's "Aphorisms" that contributed to his acclaim. Foreseeing the unity of science he declares "that a science be constituted which may be a receptacle for all axioms which are not particular to any of the individual sciences, but belong to several of them in common" (returning to Warhaft, p. 412). Continuing, he asserts that "Nature to be commanded must be obeyed" (p. 331); "that Time is like a river, which has brought down to us things light and puffed up, while those which are weighty and solid have sunk" (p. 305); and that "a like judgment I suppose may be passed on myself in future ages: that I did no great things, but simply made less account of things that were accounted great." (p. 361) This latter is highly commendable for its insight, frankness, and humility for he *was* much greater as an advocate than as an achiever.

One of his best known aphorisms is the characterization of current scientists by the ant, the spider, and the bee.

> Those who have handled sciences have been either men of experiment or men of dogmas. The men of experiment are like the ant: they only collect and use; the reasoners resemble spiders, make cobwebs out of their own substance. But the bee takes a middle course: it gathers its material from the flowers of the garden and the field but transforms and digest it by a power of its own. (p. 360)

He then continues by making his famous comparison between these various insects and current scientists.

> Not unlike this is the true business of philosophy [science], for it neither relies solely or chiefly on the powers of the mind, nor does it take the matter which it gathers from natural history and mechanical experiments and lay it up in the memory whole as it finds it, but lays it up in the understand altered and digested. Therefore from a closer

and purer league between these two faculties, the experimental and the rational (such as has never yet been made), much may be hoped. (p. 360; brackets added)

Here again he is entirely unaware that each of the contributors to modern science previously described would exemplify the bee, their achievements necessarily combining both experiments and the use of instruments such as the terrella or telescopic with mathematical correlations and theoretical reasoning. The irony, in my opinion, is that while Bacon was able to express so perceptively with his Idols the natural human impediments to obtaining true knowledge, along with the defects of traditional methodologies and current scientific knowledge, his own *Instauration* and *New Organon*, except for some astute Aphorisms, lacked a total understanding of the method needed to advance science in the manner he ardently sought.

As I have often repeated, unlike Gilbert, Kepler, Galileo, and Harvey, he was unable to extricate himself from the mythical religious beliefs of the time, still adhering to the Genesis account of Adam's fall and belief that explaining the "configuration of objects" involved discovering their "inner spirit," rather than seeking empirical or natural explanations. Though histories of science may extol Bacon's exhortations for a new science, they cannot refer to a single actual contribution by him. Thus he resembles a physician who can skillfully diagnose the symptoms of a malady, but has no conception of how to cure it!

Turning to Descartes, he in contrast was not just an ardent advocate of a new methodology, he was a multifaceted genius who pursued research in many areas (with varying success), including mathematics, psychology, medicine, physiology, embryology, dioptrics, optics, physics, and astronomy. Famous as the father of analytic geometry and the creator of Cartesian coordinates still in use, and for introducing negative square roots in algebra along with other notation, mathematics was his initial area of study in which he made his most original contributions and apparently was the model for the method of inquiry and conception of knowledge that he later devised.

But despite these significant achievements, the reason for discussing him is not because of his scientific investigations that for the most part were not outstanding or lasting, but because they illustrate how someone as brilliant as Descartes was unable to accept the methodology of science being created and implemented by the contemporary scientists. Instead, he introduced "the method of doubt" for discerning clear and distinct ideas (as in mathematics) as the foundation of his system whose truth was guaranteed by God. But though his general conception of the foundation of scientific knowledge was mistaken, he unfortunately was successful in redirecting *philosophy* from the objective approach of Aristotle and the founders of modern science (who were called "natural philosophers") to a subjectivist orientation in which the self-evidence of ideas, rather than their empirical veracity, were their test of truth and the basis of knowledge. This led modern philosophers to pursue metaphysical speculations and systems, rather than empirical inquiry, during three and a half centuries before being eclipsed by science.[44] There are so many excellent biographies of Descartes that I shall relate only the important facts that contribute to an understanding of his character, intellectual development, and later endeavors.[45]

He was born in La Haye, Touraine on March 31, 1596, about a half century after Copernicus published his *De Revolutionibus* (1543) and the birth of Gilbert (1540), but dying in 1650 he outlived Kepler by twenty years and Galileo by eight, dying eight years after Newton was born. There is much controversy among his early biographers regarding his ancestral background, two asserting (without providing evidence) that his family could claim noble lineage. Like Galileo, his ancestry did include a great grandfather who was an illustrious doctor, though unlike Galileo whose father wanted him to become a physician to restore the family's reputation, Descartes' father, Joachim Descartes, insisted "that his sons and grandsons pursue legal careers and the purchase of practices" so the family could acquire the status of nobility 'indefeasibly in the third generation.'" (p. 2) But this was not to be Descartes' destiny.

His mother died while giving birth to another son less than a year after giving birth to him, and because the second son died three days after his mother but Descartes was never told of this, he went through life believing his birth was the cause of her death. Apparently he was raised by his father and a nurse. Though there is disagreement about the dates, he was educated at the Jesuit College of La Flèche where he studied with his brother Pierre for about eight years. The purpose of the curriculum was mainly to prepare students to become teachers or professors, the emphasis being on philosophy which included logic, physics, and metaphysics.

He also might have had some preparation in mathematics because when he met Isaac Beeckman in 1618 he evinced some knowledge of the subject. Among additional courses were grammar, rhetoric, ethics, and Greek, along with eight years of Latin that some biographers claim accounts for the clarity of his excellent writing style, although this also could be attributed to his great love, from an early age, of poetry and prose. Because of poor health he was given special treatment at La Flèche, including a private room where he could remain in bed later in the morning than the other students who had to rise at five o'clock, but gradually overcame his weaknesses enabling him later to enlist in the army.

After graduating from La Flèche he followed his father's injunction to pursue a law degree at the University of Poitiers where he studied jurisprudence during the academic year 1615-16, passing with honors two consecutive examinations for the *baccalauréat* and the *licence*. Although he does not appear to have practiced law, he nonetheless kept his right to do so as "'councillor of the king at the presidial of Poitiers' until 1621." (p. 22) Then, having become of age in March 1618 (and the different accounts by his biographers make the history rather sketchy), he was free to join the army as a volunteer despite his father's forebodings that because he was not of sufficient noble status he would not attain a high rank. Deciding not to enter the king's service, he went to Breda in the Netherlands to enlist in the army of Maurice of Nassau who became Prince of Orange.

He seems to have been attracted to the military not to engage in warfare, but to be able to travel, to be a "spectator rather than an actor," as he said. According to Rodis-Lewis, after three years of war against the Turks "'in and around Hungary,' there were also journeys 'to learn virtue in the school of life, seeing various countries and various nations.'" (p. 20) His relatively short life of 54 years was spent mainly traveling and writing with several sojourns in Royal Courts. What must seem strange to anyone who has served in the military in modern times, as a volunteer in the army he was not paid for his service and thus was free to travel wherever his fancy took him, staying where he pleased and occupying himself with whatever interested him. Even when the army was at war he apparently had no specific military duties.

This seems to have been the purpose of his serving, not to participate in wars but to gain as much experience of the world as possible from observing its different languages, customs, forms of government, and wars. As examples of this diversity and his versatility, he observed the Turkish war in Hungary and the beginning of the Thirty Years War in Germany, along with witnessing the coronation celebrations when Ferdinand of Hapsburg was crowned emperor in Frankfort in 1619. (Cf. p. 33) He even had a life threatening encounter when several sailors who, thinking he was a defenseless merchant, attacked him and his valet intending to steal his money and possibly kill them. However, unprepared for his being an accomplished swordsman, he thwarted their attack when he wielded his sword and disarmed them. He also fathered an illegitimate child and fought a dual over the honor of a woman, but spared the life of the man he disarmed.

During this period he corresponded with and visited many of the leading scholars in Europe, such as Beeckman and Mersenne, although he never consulted Galileo to learn whether any of the latter's views could settle the doubts that constantly plagued him and were the motivation for his numerous and extremely influential publications. These extensive voyages and personal encounters, comprehensively described by Rodis-Lewis in her book, filled the decade between his graduation from the University of Poitiers and joining the army in 1618 to his first written work in 1628, *Rules for the Direction of the Mind*, which was unfinished and only published posthumously in Amsterdam in 1701.[46]

Deciding that none of the astute minds he had met had been able to formulate with certainty a philosophy that would withstand his doubts, like Bacon he decided to make the attempt himself which he did in a more explicit and less literary or flamboyant fashion. Although intended to include thirty-six Rules, the treatise published after his death included only twenty-one. It is divided into two parts, the first providing a general account of his new method and the second its application to mathematics. As only the first pertains to his philosophy I shall limit the discussion to it.

Stating in Rule I that its purpose was "*to direct the mind towards the enunciation of sound and correct judgments in all matter that come before it*," in Rule II he asserts the following principle: "Thus in accordance with the above maxim we reject all such merely probable knowledge and make it a rule to trust only what is completely known and incapable of being doubted." (p. 3) This of course meant rejecting all scientific knowledge. Rule III states that there are just two mental operations that provide this indubitable knowledge: intuition and deduction. Intuition he defines as "the undoubting conception of an unclouded and attentive mind, and springs from the light of reason alone; it is more certain than deduction itself, in that it is simpler...." (p. 7) In contrast deduction, as in mathematics and logic, is understood as "all necessary inference from other facts that are known with certainty...deduced from true and known principles by the continuous and uninterrupted action of a mind that has a clear vision of each step in the process." (p. 8) Being the only secure methods of attaining certainty, the mind should admit no others.

In Rule IV he offers arithmetic and geometry as instances of this kind of certainty, along with the new algebraic method. (Cf. p. 10) Rule VIII claims that "while it is the understanding alone which is capable of knowing, it yet is either helped or hindered by three other faculties, namely imagination, sense, and memory," thus he examines these to determine "where each may prove to be an impediment...or where it may profit us...." (p. 27) Rule IX emphasizes the importance of beginning investigations with simple facts before attempting an understanding of more complex ones, declaring that everyone should realize that "none of the sciences, however abstruse, is to be deduced from lofty and obscure matters, but that they all proceed only from what is easy and more readily understood." (p. 29)

Like Bacon, in Rule X he criticizes dialectics, the form of reasoning favored by the scholastics without mentioning them, arguing that it is circular because "Dialecticians are unable to devise any syllogism which has a true conclusion, unless they have first secured the material out of which to construct it; i.e., unless they have already ascertained the very truth which is deduced in that syllogism." (p. 32) In Rule XI he expounds on the two aspects comprising certain knowledge, intuition and deduction. (p. 33) Rule XII describes the roles of the external senses, common sense, imagination, fancy, and memory in forming our ordinary perceptions and conceptions of the world, concluding that the "power by which we are properly said to know things is purely spiritual...and distinct from every part of the body...." (p. 38)

He next examines the interrelation of concepts or propositions which can be either "necessarily or contingently conjoined," the union necessary when we cannot conceive either as distinct, but contingent when "conjoined by no inseparable bond." (pp. 42–43) His examples of necessarily connected propositions are particularly significant because they will exemplify his foundation of knowledge; e.g.: "'I exist, therefore God exists' and 'I know, therefore I have a mind distinct from my body' etc." (p. 43) Such propositions "do not arise as a result of inquiry, but present themselves to us spontaneously" according to the first twelve rules, "in which, we believe, we have displayed everything which, in our opinion, can facilitate the exercise of our reason." (p. 48) How he could have been convinced that these propositions can be the indubitable foundation of a universal theory of knowledge, rather than just his own idiosyncratic system of beliefs, is astonishing for a person of his critical intelligence.

The next work to be considered, *Discourse on Method and Essays*, was his first published work when he was forty-two years old in 1636. It begins again by reviewing his intellectual development and avowing the complete dedication of his life to seeking the truth. But rather than being guided by the multiplicity of Rules previously specified, his *Method* now condenses them into just four precepts or conjunctions of reasoning presented in Part II: first, "to accept nothing as true which I did not clearly recognize to be so;" second, "to divide up each of the difficulties which I examined into as many parts as possible...in order that it might be resolved in the best manner possible;" third, "to carry on my reflections in due order, commencing with objects that were the most simple and easy to understand, in order to rise little by little, or by degrees, to knowledge of the most complex;" and fourth, "in all cases to make enumerations so complete and reviews so general that I should be certain of having omitted nothing." (p. 92) He adds that "what pleased me most in this Method was that I was certain by its means of exercising my reason in all things, if not perfectly, at least as well as was in my power." (p. 94)

It is in Part IV that we come to his crucial argument that is repeated in his later works. His method, as we know, is to start by rejecting as "absolutely false everything as to which I could imagine the least ground of doubt, in order to see if afterwards there remained anything in my belief that was entirely certain." (p. 101) The rejected knowledge includes sensory perceptions because our senses sometimes deceive us; demonstrations because they can lead to "paralogisms," even in geometry; and finally, since whatever occupies our consciousness while awake may also come to us in sleep, we can doubt that any of its contents has anymore truth than dreams. (Cf. 101) This leads to his famous inference that even though he can doubt the veridicality of other contents of consciousness, he cannot doubt that he must exist to be in a state of doubt, hence the conclusion "*I think therefore I am.*"

> But immediately afterwards I noticed that whilst I thus wished to think all things false, it was absolutely essential that the 'I' who thought this should be somewhat, and remarking that this truth '*I think, therefore I am*' was so certain and so assured that all the most extravagant suppositions brought forward by the skeptics were incapable of shaking it, I came to the conclusion that I could receive it without scruple as the first principle of the Philosophy for which I was seeking. (p. 101)

While we can grant that *this* proposition is indubitable, it is the inferences that he derives from it, such as that he is a "thinking soul-substance" that can exist "entirely distinct from [his] body," that it "is even more easy to know than is the latter," and that it can exist independently of it, that surely can be doubted.

> From that I knew that I was a substance the whole essence or nature of which is to think, and that for its existence there is no need of any place, nor does it depend on any material thing; so that this 'me,' that is to say, the soul by which I am what I am, is entirely distinct from body, and is even more easy to know than is the latter; and even if body were not, the soul would not cease to be what it is. (p. 101)

The crucial deduction is his inference from his knowing that he is thinking to the conclusion that he is a soul-substance that is distinct from his body and thus could continue to exist without it. As he claimed previously in Rule XII, knowledge that is certain depends upon an initial *indubitable intuition* and then *necessary deductions* from it resulting in a system of propositions which are so connected as not to be conceivably distinct and thus certain. Thus it is not the indubitable initial intuition that if he is thinking he must exist that is dubious, but that this incontrovertibly implies he must be an immaterial, spaceless soul, *which he simply equates with the 'me,'* that can exist independently of the body and the brain. It is not a question of whether it is impossible to hold the belief, but whether it is so certain as to be incontrovertible.

The argument raises all sorts of questions. In equating the soul with himself he begs the question. A strong *cognitive association* is not an indubitable *conceptual connection*! What kind of immaterial substance is thinking and how does it sustain thinking? He claims it is a soul-substance, but I doubt if even Descartes had a clear and distinct idea of what a soul-substance is. His doubt leads to a logical cul-de-sac wherein he is nothing but a spaceless, immaterial, thinking existent, which someone not already indoctrinated with the belief would hardly find intelligible, never mind indubitable!

That *he* found the argument indubitable is no guarantee of its truth, only that it's self-evidence is a consequence of his own particular background of beliefs. Is this not true of everyone, however absurd the belief? Think of the prevalent belief among Christians of the Virgin Mary and yet it is not even credible that Jesus could have been born of a Virgin despite the fact that one can affirm it. While throughout the past consciousness usually *was* attributed to the soul, the physiological and anatomical investigations of Vesalius (1514-1564) and Fabricius (1537-1619), along with Descartes' own physiological inquiries, had indicated that the ultimate cause of our actions and thoughts is our brain—which should have obviated his drawing such an implausible independence of the soul from the body. That the inference is implausible will be shown in his later extraordinary reaffirmation of all the commonsense beliefs that he has just dismissed.

As should be clear, the essential weakness in his method is that *subjective certainty is no guarantee of truth* (although it is still a prevalent criterion among religionists for accepting their beliefs because they were instilled at an early age before subjected to any questioning and thus can seem incontestable). But given Descartes' diversified experience and broad intellectual background, it is difficult to understand how he could have held that the certainty of his initial self-evident intuitions and indubitable inferences were derived from "the light of reason alone," endowed by God, as stated in Rule III. Given the range of beliefs he must have encountered in his extensive travels, it should have been obvious to him that what people consider certain is relative to their intellectual and cultural backgrounds. Yet he insisted in believing that his conclusions were unconditionally and universally true, not just true for him.

> And having remarked that there was nothing at all in the statement '*I* think, therefore I am' which assures me of having thereby made a true assertion, excepting that I see very clearly that to think it is necessary to be, I came to the conclusion that I might assume, as a general rule, that the things which we conceive very clearly and distinctly are all true.... (p. 102)

But this statement inadvertently reveals what is mistaken in the argument in that while '*I think therefore I am*' does imply "*that to think it is necessary to be,*" it *does not imply* that he is a soul-substance. What it is to be a thinking being is not implied in the assertion, but has to be inferred or explained. Unlike the statement "I think therefore I am," there is no *in*controvertible connection between the statements that "I am a thinking being" and that "I am a soul-substance" indicated by the fact that they can be denied: "I am a thinking being but I am not a soul-substance," thus violating his definition of what is an indubitable connection between statements.

Thus the issue is not just ascertaining which ideas are initially indubitable, but what one is justified in inferring from them! Both the initial premises and implied inferences depend on one's background beliefs which is true even of most of the ideas that we consider self-evident, except for those that are tautologies which "I think, therefore I am" resembles. The classic example was the parallel axiom in Euclidean geometry that only one straight line in the plane through a given point outside the line can be drawn parallel to it that was considered to be self-evidently true throughout most of the past. However, when non-Euclidean geometries were developed in the 19th century showing that any number of lines can intersect through a given point in curved space, the Euclidean axiom became merely a postulate relative to whether space is flat or curved. Quantum mechanics is another example of the rejection of many intuitive certainties of classical physics.

The flaw in Descartes' reasoning becomes even more apparent when we consider additional arguments in which he claims to draw other necessary conclusions from self-evident ideas. Declaring that since doubting is an imperfection and yet he has the idea of something more perfect than himself, he recognized "very clearly" that this conception must proceed from some nature which was more perfect. Thus he infered that "the idea of a Being more perfect than my own…had been placed in me by a Nature which was really more perfect than mine could be, and which even had within itself all the perfections of which I could form any idea—that is to say, to put it in a word, which was God." (p. 102)

While this may have been self-evident to Descartes, I doubt that many others would be convinced that the idea of perfection required a God as its cause, even if they believe in God. Given all the gradations in the world, the idea of something uniquely perfect could more simply be explained as derived from this realization, rather then having been endowed by God. Furthermore, had the idea of God been implanted in him, then unless he was privileged everyone should have had the same idea of and belief in God, which is far from true. Most deities of the past were far from perfect.

Though Descartes argues that to be convinced of his reasoning it was necessary to free his thinking from the influence of the senses to think rationally, I believe a better explanation of his believing with certainty the existence of a Perfect Being is found in his frank admission in the first maxim in Part III, that his doubt would be guided by "adhering constantly to the religion in which by God's grace I had been instructed since my childhood.…" (p. 95) While I know from what a friend has told me that his experience of God's grace is so overwhelming that he could not possibly doubt its authenticity, yet the certitude of subjective experiences has never been a guarantee of their truth. Consider all the beliefs once held by humankind which at the time seemed incontrovertible, such as the existence of a heaven and hell or that committing certain sins condemned one to purgatory. The history of science too is replete with discarded ideas which previously were thought to be self-evidently true. Which, then, is more plausible, that the existence of God is demonstrated by his having the idea of a Perfect Being or that he acquired the belief when he was a child, as he claimed?

The argument is reinforced by his declaring that the basis for distinguishing between the illusions of dreams and the reality of waking experience is the belief that God, because of his goodness, would not deceive us.

> For how do we know that the thoughts that come in dreams are more false than those that we have when we are a wake, seeing that often…the former are not less lively and vivid than the latter? And though the wisest minds may study the matter as much as they will, I do not believe that they will be able to give any sufficient reason for removing this doubt, unless they presuppose the existence of God. For…all the things that we very clearly and very distinctly conceive are true is certain only because God is or exists, and that He is a Perfect Being.… (p. 105)

Once again the argument is unconvincing. Imagine waking up from a terrifying dream and being momentarily unable to decided which is actual, the dream or one's awakened state, and deciding in terms of God's probity? We know we are not dreaming when we awake and find ourselves in bed relieved that the nightmare is over. It is true that some dreams are so realistic that for a short time after awakening we are still beset by the dream, but is the awareness that it was just a dream due to the realization that God would not deceive us or to the fact that our dreams are not coherent with our normal waking life? Here again the conviction is not cognitively compelling. Even Descartes, in his later writings, will introduce the discontinuity as the distinguishing factor, but it is puzzling how he could have believed it earlier. Nor is it true that "all the things that we very clearly and very distinctly conceive are true…only because God is or exists," considering the doubtfulness of God's existence and how many seemingly self-evident truths of the past turned out to be false.

His further unlikely claim is that natural laws are "imprinted on the mind" despite the difficulty and delay in discovering them. If the laws of planetary motion were imprinted on the mind, why did Kepler have to rely on Tycho Brahe's astronomical data to decide that Mars's orbit is elliptical and then formulate the law that the periodic times squared are proportional to the mean distances from the sun cubed? Or why did Galileo have to correct Aristotle's claim that the acceleration of falling objects is proportional to their weights by his incline plane experiments? Yet Descartes declares,

> I have also observed certain laws which God has so established in Nature, and of which He has imprinted such ideas in our minds, that, after having reflected sufficiently upon the matter we cannot doubt their being accurately observed in all that exists or is done in the world. Further, in considering the sequence of these laws, it seems to me that

I have discovered many truths more useful and more important than all that I have formerly learned or even hoped to learn. (pp. 106–107)

If he had discovered such laws and truths he certainly kept them well hidden because there is no record of them. He is credited with formulating two laws, the law of inertia which was a generalization of Galileo's restricted law and the law of optical refraction which is also attributed to Willibrord Snell.[47] Although he did extensive scientific research in medicine, dissection, dioptrics, and meteors, this did not result in the discovery of any recorded laws that I know of. And while it is true that some outstanding scientists, such as Newton and Einstein (and previously the well-known contemporary cosmologist Steven Hawking) believed that the laws they discovered *preexisted in God's mind*, their discoveries were not based on introspective intuitions guaranteed by God's veracity, but on observational or experimental evidence. None would have made the marvelous discoveries they did had they followed the methodollogy of Descartes.

Consider also all the laws or beliefs in the past that were considered self-evidently true and which have since been disproved: that the earth is motionless and stationary in the center of the universe; that the planets move in circular orbits with uniform motion; that the cosmos is spherical and finite; that the velocity of light is instantaneous; that species are preformed and immutable; and that evolution requires intelligent design. But even these accepted ideas were based initially on observations. Can one think of any law that was derived just by introspection?

The remainder of the *Discourse on Method* contains a summary of the contents of his previously unpublished Treatise, *Le Monde*, which had been "suppressed or destroyed on his hearing of the condemnation of Galileo in 1632." (p. 80) But since he will describe his scientific investigations in later works, such as *The Principles of Philosophy* to be considered shortly, they will not be discussed now. Instead, I will turn to his most famous treatise, the long title of which is *Meditations On The First Philosophy In Which The Existence Of God And The Distinction Between Mind And Body Are Demonstrated*, but which fortunately has been abbreviated to either *Meditations On First Philosophy* or simply *Meditations*. The first Latin edition was published in Paris by Michael Soly in 1641 by Royal "privilege" and a second edition in French translated by Duc de Luynes published in Amsterdam by Louis Elzevir in 1642, the publisher who had printed Galileo's *Dialogues Concerning Two New Sciences*.

Descartes' initial arguments in the *Meditations* are practically the same as in the *Discourse on Method*, only more refined. In the *Meditations* he argues that since our senses are known to be deceptive he is justified in doubting their veracity (the controversy over the Copernican revolution undoubtedly contributing to this). He then presents the well-known image of his sitting before a fire in his dressing gown holding a sheet of paper and wondering whether he could really doubt that he, with his body and hands, was truly seated there, but justifies the doubt by recalling that deranged people are capable of believing all sorts of incredible things (which is why they are considered deranged).

In addition, he again evokes dreams so lifelike that he wonders whether even his normal life is nothing but a dream, drawing the well-known conclusion that while he cannot doubt that he is having the experience he is having, whether dreaming or awake, he can doubt whether what his experience *represents* is true or exists. "Now as to what concerns ideas, if we consider them only in themselves and do not relate them to anything else beyond themselves, they cannot properly speaking be false...." (p. 159) It is this detachment of the contents of his consciousness, like a painting, mirror image, or photograph from what it represents, that was crucial to his argument and that formed the essential epistemological problem of modern philosophy, as to what degree our perceptions (or concepts and theories) truly represents the real world, if at all.

Even then he wonders whether God in all his majesty could not have created him with the seemingly veridical sensory experiences he has, along with such indubitable knowledge as $2 + 3 = 5$, though they were actually illusions. But realizing that this deception would be inconsistent with God's goodness, he conjures the image of an evil genius who *does* deliberately deceive him and wonders if there were any way he could extricate himself from the deception.

> I shall then suppose, not the God who is supremely good and the fountain of truth, but some evil genius not less powerful than deceitful, has employed his whole energies in deceiving me; I shall consider that the heavens, the earth, colours, figures, sound, and all other external things are naught but the illusions and dreams of which this genius has availed himself in order to lay traps for my credulity.... (p. 148)

It is in *Meditation* II that he again presents his rebuttal, for though he can doubt that all his experiences and thoughts are truthful, he cannot doubt that he at least must exist to do the doubting, again coming "to the definite conclusion that this proposition: I am, I exist, is necessarily true each time that I pronounce it, or that I mentally conceive it." (p. 150) Reaffirming that he is "A thing which thinks," he asks, "What is a thing which thinks?", concluding that "It is a thing which doubts, understands, affirms, denies, wills, refuses, which also imagines and feels." (p. 153). But though he identifies this "thing which thinks" with a mind or soul or reason, he has not explained what the latter are or how they exist now that he has dismissed his body as their source.

This explanation is presented in *Meditation* III where he asserts that "the principal error...consists in my judging that the ideas which are in me are similar or conformable to the things which are outside me...." (p. 160). He then reintroduces the contrast between "adventitious" and "innate ideas " based on what we assume caused them, our belief in the former arising from "spontaneous inclination" while the latter are due to "a natural light which makes me recognize that it is true." (p. 160) He then introduces a second criterion of the truth of an idea, in addition to its intuitive self-evidence, based on the "objective reality" implied in the idea and its efficient cause.

Declaring that it is manifest by a "natural light that there must be at least as much reality in the efficient and total cause as in its effect," he concludes again that an imperfect being like himself could not be the cause of the idea of God that contains so much inherent perfection, hence God must exist to cause him to have the idea. He repeats the conclusion that "one certainly ought not to find it strange that God, in creating me, placed this idea within me to be like the mark of the workman imprinted on his work...." (p. 170) This conviction is repeated today in the form that the belief in God is "hardwired in us," despite the fact that there have always been dissenters and that their number is growing, at least in Europe.

Then in *Meditation* VII, the final one, apparently reflecting the many *Objections* and his *Replies*, he incredibly reaffirms his belief in all the things he had previously doubted, evincing his considerable knowledge of anatomy and physiology in returning to a more naturalistic and reasonable view of experience, knowledge, and existence. Having previously detached his perceptions, sensations, feelings, and ideas from the external world and disavowed his body and brain as their causes, he now reinstates them.

Convinced of the existence of God, one would have expected that he would adopt the later argument of Bishop Berkeley that *esse est percipi* (to be is to be perceived), and simply claim that God *directly* causes these experiences in us. However, he rejects this explanation on the grounds that "since God is no deceiver, it is very manifest that He does not communicate to me these ideas immediately and by Himself...." (p. 191) Faced with the choice either that God is the sole cause of our experiences of the world or that nature is the cause of these experiences, he now chooses the latter.

> For since He has given me...a very great inclination to believe that they [his experiences] are conveyed to me by corporeal objects, I do not see how He could be defended from the accusation of deceit if these ideas were produced by causes other than corporeal objects. Hence we must allow that corporeal things exist. (p. 191; brackets added)

So he now maintains that our bodies and brains are the mediate or approximate causes of our experiences and thus united to us: "there is nothing which this nature teaches me more expressly than that I have a body which is adversely affected when I feel pain, which has need of food or drink when I experience the feelings of hunger and thirst, and so on; nor can I doubt there being some truth in all this." (p. 192) He thus restores the explanation that nature is the source of these bodily experiences.

> Nature also teaches me by these sensations...that I am not only lodged in my body as a pilot in a vessel, but that I am very closely united to it.... Moreover, nature teaches me that many other bodies exist around mine, of which some are to be avoided, and others sought after. And certainly from the fact that I am sensible of different sorts of colours, sounds, scents, tastes, heat, hardness, etc., I very easily conclude that there are in the bodies from which all these diverse sense-perceptions proceed certain variations which answer to them, although possibly these are not really at all similar to them. (p. 192)

He declares his agreement with Harvey's circulation of the blood, concedes the dependence of mental processes on the brain, even affirming the modern neurological discovery that particular experiences and bodily functions are localized in the brain.

> I further notice that the mind does not receive the impressions from all parts of the body immediately, *but only from the brain*, or perhaps even from one of its smallest parts...in which the common sense is said to reside, which, whenever it is disposed in the same particular way, conveys the same thing to the mind, although meanwhile the other portions of the body may be differently disposed, as is testified by innumerable experiments.... (p. 196; italics added)

This paradoxically exemplifies the sophistication of his experimental inquiries compared with his previous arguments. That his knowledge of neurological processes exceeded that of Aristotle is shown by his locating the *sensus communis* in the brain, rather than in the heart, and his explanation of sensory deceptions, such as the phantom limb, as due to impaired nerve connections to or from the brain. He also ridicules and rejects his previous doubts, especially those due to dreaming.

> And I ought to set aside all the doubts of these past days as...ridiculous, particularly that very common uncertainty respecting sleep, which I could not distinguish from the waking state; for at present I find a very notable difference between the two, inasmuch as our memory can never connect our dreams one with the other, or with the whole course of our lives, as it unites events...while we are awake. (pp. 198–199)

Thus he corrects his earlier argument that dreams, being indistinguishable at times from waking experiences, is a justification for severing all experiences from the body and attributing them to an independent mind or soul, the position I formerly criticized. He now claims, as I argued, that they can be distinguished because dreams cannot be integrated with our waking experiences, thus refuting his former dualism of the mind and body. It is amazing to find how easily he can dismiss his earlier doubts and return to a more common sense view of the world, though he should be commened for addressing the arguments of friends and critics that provoked these revisions.

Turning next to his *Principles of Philosophy*, as indicated previously the edition was published in Latin by the Elzevirs in 1644 and later translated into French by his friend Picot but apparently not published until 1647, three years before his death, since that is the year he wrote a "Letter to the Translator" which serves as a Preface to the book. In it he presents his conception of philosophy as the pursuit of wisdom, by which he means not only that he understands "prudence in affairs, but also *a perfect knowledge of all the things that man can know*, both for the conduct of his life and for the conservation of his health and the invention of all the arts," a rather extravagant claim. (pp. 203–204; italics added) Asserting that not only does philosophy extend over the whole range of human knowledge (which science does today), but that "it alone is what distinguishes us from savages and barbarians, and that the civilization and refinement of each nation is proportionate to the superiority of its philosophy." (p. 204) He then explains "in what all the knowledge we now possess consists, and to what degrees of wisdom we have attained." (p. 205)

> The first of these contains only notions which are…so clear that they may be acquired without any meditation. The second comprehends all that which the experience of the senses shows us. The third, what the conversation of other men teaches us. And for the fourth we may add to this the reading, not of all books, but especially of those which have been written by persons who are capable of conveying good instruction to us, for this is a species of conversation held with their authors. And it seems to me that all the wisdom that we usually possess is acquired by these four means only; for I do not place divine revelation in the same rank, because it does not lead us by degrees, but raises us at a stroke to an infallible belief. (pp. 205–206)

Once again he describes the process by which, starting from doubt, he was able to resolve his doubts confident that God, the author of the world, has endowed him with an understanding capable of arriving at the "principles" (sounding like Bacon) of which "I make use respecting immaterial or metaphysical things, from which I very clearly deduce those of corporeal or physical things…." (p. 208). Displaying a marked hubris he declares that though not having considered everything, he has

> so explained all those matters with which I have had occasion to deal, that those who read them with attention will have reason to persuade themselves that there is no need to seek other principles than those I have brought forward, in order to arrive at all the most exalted knowledge of which the human mind is capable. (p. 209)

He implies that those who do not agree with his reasoning do so because they are not sufficiently attentive, though at other times he attributes this to prejudice or to their not having freed their thinking from their senses. Interestingly, though Galileo and Newton contributed enormously to the creation of modern science they were very modest in appraising their achievements, while Bacon and Descartes, who failed to make any actual contribution to the advancement of science lavishly praised the prospects of their alternative misguided methodologies.

As the following quotation indicates, by applying his "principles" Descartes claims to have been able to "expound" the whole range of knowledge.

> The principles producing this knowledge and wisdom are divided into four parts, the first of which contains the principles of knowledge, which is what may be called the First Philosophy or Metaphysics …. The other three parts contain all that is most general in Physics, i.e. an explanation of the first laws or principles of nature, the manner in which the heavens and fixed stars, the planets, the comets, and generally all [of which] the universe is composed. Then the nature of this earth, and of the air, water, fire, and the loadstone, is dealt with more particularly…as also all the quality as observed in these bodies, such as light, heat, weight, and such like. By this means I believe myself to have commenced to expound the whole of philosophy…. (p. 212; brackets added)

What an extraordinary display of bravura!

For those willing to examine the 574 *Principles of Philosophy* and the *Principles of Human Knowledge*, along with their expository comments, they will find a tremendous range of knowledge including some correct laws and judgments regarding current theories, but also many misconceptions, errors, and even contradictions that is to be expected in such a broad undertaking. They reflect not only Descartes' understanding (or rather misunderstanding) of the science of his day, but also the still limited and confused state of knowledge then existing. Foregoing such an extensive critique, I will conclude with an overall evaluation of his philosophy.

Descartes lived when modern classical science was just being formed, a period of extensive scientific inquiries and developments owing to the increasing use of experimentation, the construction of instruments like the terrella, telescope, and microscope, and the introduction of perplexing explanations and radical new theories. In contention was the very structure of the universe owing to the introduction of a new cosmological system starting with Copernicus's hypothesis, Kepler's astronomical laws, and Galileo's telescopic discoveries. Not only displacing human beings from the center of the universe, they were undermining two main traditional sources of knowledge at the time, the Bible that included the creation account in Genesis and the Aristotelian geocentric conception of the universe and theory of scientific explanation. They also involved the beginning of a new understanding of the nature of humankind. As previously mentioned, experiments and discoveries in biology, anatomy, and physiology by Vesalius, Fabricius, and Harvey describing the circulation of the blood and the function of the nerves and the brain were transforming the conception of human beings, challenging the soul as the immortal seat of consciousness. Unlike the scientists who also had a firm belief in God but were convinced that to be understood nature had to be investigated objectively aided by observation, inductive reasoning, experimentation, and the discovery of mathematical laws and despite his own experimental investigations in optics and neurology, Descartes still proposed a philosophical method of inquiry that basically relied on clear and distinct ideas whose truth was guaranteed by the light of reason endowed by God and whose formulation was metaphysics. Though finally realizing that his method of doubt lead to absurd beliefs, he still claimed that by following "his laws of nature and principles of philosophy" he was able to explain the whole range of knowledge. But Newton, after an acute study of Descartes' works, in his "Quaestiones" published in 1664, severely criticized his theories of light, colors, gravity, the tides, and vortices..

But as the father of modern philosophy his main legacy was to initiate a new method of inquiry distinct from *natural* philosophy (science) founded on subjective ideas constituting metaphysical knowledge, complementing empirical inquiry. Thus he asserted that after forming a "code of morals" to guide one's life, one studies mathematics and logic to improve one's thinking; metaphysics to acquire the "principles of knowledge" which explain "the principal attributes of God, of the immateriality of our souls, and of all the clear and simple notions that are within us;" physics to comprehend the "true principles of material things" to explain how the whole universe is composed; and finally the entire range of empirical knowledge. (Cf. p. 211). As he wrote in a famous passage: "Thus philosophy as a whole is like a tree whose roots are metaphysics, whose trunk is physics, and whose branches, which issue from this trunk, are all the other sciences." (p. 211)

The effect of grounding knowledge on the clarity of ideas and metaphysics, rather than scientific inquiry, led succeeding generations of philosophers to a subjectivist conception of knowledge that claimed to supplement or replace scientific explanations with metaphysical systems.[48] That led to Berkeley's *esse est percipi* replacing the physical world with a divine being as the direct cause of experience and creation; Hume's restriction of all knowledge to impressions and ideas and consequent denial of discovering any underlying causal relations in nature; Kant's belief that the world as experienced is the phenomenal world constructed entirely by our minds, though somehow influenced by unknowable things-in-themselves; Hegel's *Phenomenology of Mind* that described the world as the dialectical manifestation of an Absolute Spirit; Schopenhauer's *The World as Will and Representation*; and F. H. Bradley's *Appearance and Reality*, etc. Each of these metaphysical systems are now obsolete, scientific inquiry refuting the contention that the world can be understood solely as a manifestation of the mind, not as an investigation of nature—though somewhat mitigated by the uncertainties of quantum mechanics.

Thus the achievements of science since Descartes' time stand in marked contrast to philosophy. Not what Descartes anticipated, it proved to be the consequence of his failed attempt to devise a methodology more fundamental than scientific inquiry based on "clear and distinct ideas and necessary inferences." What he considered to be the foundation or root of all knowledge, metaphysics as "the queen of the sciences" as it was once known, hardly exists as a discipline today having been replaced by science. And since we often learn of things by way of their contrasts or mistakes, this was the reason for the discussion of Descartes, along with Bacon, in this reconstrucion of the origins of modern science and its role in enlightening modern civilization.

CHAPTER EIGHT

Newton's Incomparable Achievements

We turn finally to the venerated Isaac Newton who was able to forge the contributions of Copernicus, Kepler, Galileo, Hooke, Huygens, and Boyle into a coherent methodology, scientific framework, and cosmological worldview that finally replaced the Genesis account of creation and Aristotle's entire natural philosophical system. This amazing achievement lasted for about two centuries until quantum mechanics displaced Newtonian mechanics in subatomic investigations and Einstein's special and general theories of relativity introduced a new cosmological framework. As Newton and Einstein are considered the two greatest physicists, it is intriguing to ask which is the greater of the two before turning to a discussion of Newton's accomplishments?

In past writings I have claimed that it was Newton because of his superlative contributions to the *three* major areas of scientific inquiry: experimental based on his optical investigations and prismatic discovery that white light is not homogeneous but composed of discrete color-rays; theoretical with his formulations of the universal laws of motion and system of celestial mechanics that finally replaced Aristotle's distinction between the celestial and terrestrial worlds; and mathematical owing to his prodigious mathematical talents that enabled him solve complex mathematical problems and create fluxional or differential calculus to compute the dimensions of planetary orbits in his celestial mechanics.

Einstein also was a brilliant theoretical physicist who in 1905, when he was just twenty-six years old wrote five papers, three of which had a revolutionary impact on physics. The first was on Brownian motion in which he showed mathematically that the agitation of pollen grains suspended in water was caused by the impact of the water molecules that finally convinced most scientists of the existence of atomic-molecular particles. Two other articles were largely responsible for the overthrow of Newtonian mechanics as the fundamental or final explanation of the universe. One was on the photoelectric effect explaining that the illumination on the surface of metal plates when radiated by ultraviolet light was caused by light quanta (later called photons) striking the inner vibrating atoms with sufficient energy to eject the electrons that were the source of the illumination. Reinforcing Planck's quantum mechanics, it contributed to the particle theory of light and to the rejection of Newtonian determinism at the subatomic level.

The third article on his theory of special relativity showed that Newton's absolute frames of the universe, space and time, were only approximately true when the velocity of physical interactions was insignificant compared to the speed of light, demonstrating the relativity of measurements but the universality of the laws of nature with the velocity of light the limiting velocity. Then his general theory of relativity published in 1915, in which he proposed that Newton's gravitational force be interpreted as a curved four-dimensional space-time, supplanted Newton's cosmological framework, especially in predicting black holes despite Einstein's ironic persistent refusal to admit their existence.

But unlike Newton who was self-taught in mathematics yet became a leading mathematician of his day and created the method of fluxions to aid him in solving his astronomical calculations, Einstein occasionally relied on his friends when he needed mathematical assistance and adopted Riemann's non-Euclidean geometry to depict the gravitational force warping the structure of four-dimensional space-time. Also, he did not make any experimental discoveries, though he did rely on ingenious thought experiments (such as comparing gravitational forces in an enclosed free falling elevator with gravitational free outer space) when developing his relativity theories. Still, his mathematical proof that the speed of light in a vacuum is constant is comparable to Newton's demonstration of the finite speed of light. And just as Newton was indebted to Kepler's third law, Einstein owed much to Maxwell's theory of electromagnetic fields.

When my wife and I visited the Royal Society in London in December 2007, Keith Moore, the Head of Library and Information Services, graciously showed us around the library and portrait gallery, even producing the first edition of Newton's *Principia Mathematica* (handwritten with marginal notes in a large binding) when he asked if there was a book in their collection I would like to examine. I was so pleased by the experience, that to show my appreciation I expressed my sincere choice of Newton as the greatest scientist of all time. To my further satisfaction, Mr. Moore told me that at a recent debate by members of the Royal Society who also considered the question, they gave the vote to Newton by a small majority—although I can't help wondering what the outcome would have been if the vote had taken place at the University of Berlin. In either case, to be one of the two undisputed greatest scientists who ever lived is not to be slighted.

Turning to Newton, he was born in the family manor house in Woolsthrope, Lincolnshire on Christmas day in 1642, the year Galileo died, an auspicious coincidence. His father died three months before his birth and his mother married three years later deciding to accompany her new husband leaving Newton to live with his grandmother. After his mother's second husband died ten years later she returned to Woolsthorpe with her two children by her second marriage to rejoin Newton. The death of his father whom he never knew and the departure of his mother for ten years must have left their mark on Newton's sensitive nature, especially as he was not fond of his grandmother.

He entered the three century old Free Grammar School of King Edward VI in Grantham when he was twelve years old which offered the traditional religious studies and courses in Greek and Latin. According to Richard Westfall's excellent biographical account of Newton's life and publications, he was known as "'a sober, thinking lad'" who "'never was known scarce to play with the boys abroad.'"[49] Like Copernicus, during his early years he was an extremely sensitive, reticent, withdrawn person who shunned publicity and controversy. Yet even as a young boy he showed signs of courage, as when he was attacked one day by an older school bully and fighting back decisively beat him.

But it was his exceptional intellectual gifts that led the headmaster of the school to suggest to his maternal uncle that he take the required courses to prepare for entrance to the university, which he did. He was subsequently admitted to Trinity College, Cambridge in the summer of 1661, where his uncle had studied thirty years earlier and that was reputed to be the foremost college in England at the time. He also had to matriculate at Cambridge University to receive a degree. It was only after he retired from the university and was appointed Master of the Mint and moved to London that, thanks to his niece Catherine Barton who served as his hostess and who was much admired for her intelligence, charm, and beauty, he became active socially.

At Trinity he continued his previous studies along with Aristotle's physics and cosmology, though gradually becoming more interested in the works of Kepler, Galileo, Descartes, Gassendi, Robert Boyle, and Henry More. Because at the time Descartes' works were being discussed at Cambridge more than Galileo's, Newton began studying Descartes seriously, especially his experiments in optics and his theory of vortices. These inquiries published as "Quaestiones" in 1664 contained his criticisms of Descartes' theories of light and colors, gravity, the tides, and vortices, the latter because they did not explain eclipses nor agree with Kepler's three laws.

Largely as a result of his dissatisfaction with Descartes' explanations, he was increasingly drawn to the atomic-mechanistic interpretation of nature proposed by Gassendi and Boyle and began his mathematical studies that would establish his reputation as a brilliant mathematician. Yet his future at Trinity College was in doubt because to attain a secure position he had to receive a fellowship which at that time was not based on academic performance or promise, but on social position, patronage, and seniority. Since he did not have the latter advantages despite his uncle having attended Trinity, his situation looked bleak, producing one of the "disorders" that he experienced several times in his life. Fortunately, however, he was elected to a fellowship at Trinity in 1664 though there is insufficient evidence to show how this occurred.

The following two years from 1664-66, when he was twenty two to twenty four years of age, are designated his *"anni mirabilis"* due to his intense studies resulting in three papers that applied his newly acquired mathematical skills to the problem of motion—a precursor of Einstein's achievement previously mentioned of publishing five remarkable articles in *Annalen der Physik* when he was twenty-six years old. Westfall presents a vivid assessment of his accomplishment.

> Taken in all, the tract of October 1666 on resolving problems by motion was a virtuoso performance that would have left the mathematicians of Europe breathless in admiration, envy, and awe. As it happened, only one other mathematician in Europe, Isaac Barrow, even knew that Newton existed, and it is unlikely that in 1666 Barrow had any inkling of his accomplishment. The fact that he was unknown does not alter the other fact that the young man not yet twenty-four, without benefit of formal instruction, had become the leading mathematician of Europe. And the only one who… mattered, Newton himself, understood his position clearly enough. He had studied the acknowledged masters. He knew the limits they could not surpass. He had outstripped them all, and by far. (pp. 137–138)

Appling this burst of creativity not only to the study of mathematics, he also directed it to planetary motions as a result of his critique of Descartes' conception of vortices. As Newton famously recounted to William Stukeley, his longtime friend, it was during the plague years of 1665-1666, when he returned to his home in Woolsthorpe because of the closing of Cambridge University, that one day when he was sitting in the orchard and watching an apple fall it occurred to him that if the earth's gravity caused the apple's fall, then if extended to the moon perhaps it was the same gravitational force that produced the moon's elliptical orbit. (Since the validity of these anecdotes are often questioned, I can attest that during our visit to the Royal Society mentioned earlier, Mr. Moore kindly showed my wife and me the page in Stukeley's diary where he states that Newton told him of seeing the apple fall in the orchard, and what it suggested to him, that should confirm the account.)

This led him to consider what the magnitude of the gravitational force would have to be to cause the deviation of the moon's orbit from circular to elliptical. Knowing of Kepler's third law that a planet's periodic times squared is proportional to the 3/2th power of its mean distance from the sun, he inferred that the strength of the earth's gravitational force on the moon must also be in the same ratio. As he states,

> I began to think of gravity extending to ye orb of the Moon &...from Kepler's rule of the periodical times of the Planets being in sesquialterate [3/2th] proportion of the distances from the center of their Orbs, I deduced that the forces wch keep the Planets in their Orbs must [be] reciprocally as the squares of their distance from the centers about wch they revolve & thereby compared the force requisite to keep the Moon in her Orb with the force of gravity at the surface of the earth, & found them answer pretty nearly. All this was in the two plague years of 1665-1666. For in those days I was in the prime of my age for invention & minded mathematicks and philosophy more than at any time since. (p. 143; brackets added)

I would suggest that this quotation indicates that it was more than seeing the apple fall that inspired Newton to consider whether it was the gravitational force of the earth on the moon that was the cause of its orbital motion, but also his reading of Kepler. As the above statements assert, he knew of Kepler's third law quoted previously in the latter's *Harmonice Mundi* that "the gravitational force of the sun on Mars diminishes in the ratio of the 3/2th power of the mean distances" and also that "the force requisite to keep the Moon in her Orb [compares] with the force of gravity at the surface of the earth." In addition, Kepler declared in the *Nova* that "if the Earth ceased to attract the water of the sea, the seas would rise and flow into the Moon," recognizing a mutual gravitational force between them.

Furthermore, in his *Epitome Astronomæ Copernicanæ* Kepler generalized his previous law asserting that his three laws of motion applied to all the planets, not just Mars, along with the Moon and the satellites of Jupiter and thus antedated Newton's explanation. Whether he had read all of Kepler's book, given his "intense studies" of motion and the tract of 1666 written about the time he saw the apple fall, it is likely he would have been aware at least that it was the gravitational force of the earth that produced the moon's orbit. As Koestler states: "The *Epitome* is not an abstract of the Copernican system, but a textbook of the Keplerian system. The laws which originally referred to Mars only, are here extended to all the planets, including the moon and the satellites of Jupiter."[50] Thus I agree with Robert Small that Kepler's laws "were, indeed, the foundations of the whole theory of Newton; and it will not perhaps be thought an unjust conclusion from the consideration of them, that no person, in any age, ever soared higher than Kepler, above the common elevation of his contemporaries."[51] Newton of course was not his contemporary, nor does this detract from Newton's achievements who acknowledged that "[i]f I have seen further it is by standing on ye shoulders of Giants."[52] But it does indicate who one of those giants was!

Returning to the discussion of his progressive accomplishments, among the three papers mentioned previously was one entitled, "The Lawes of Motion," in which he presented two additional theoretical discoveries that helped form the foundation of his theory of dynamics presented later in the *Principia*. In one he replaced Descartes' description of vortices interacting by contact with his own conception that they intertact by mutually reciprocating forces. In another he introduced his "principle of the conservation of angular momentum for the first time in the history of mechanics: 'every body keeps the same real quantity of circular motion and velocity so long as tis not opposed by other bodys.'"[53] In addition, he discussed his unique experiments that resulted in what he referred to as the "celebrated Phenomena of Colours."

Having studied the theories of color by Descartes, Boyle, and Hooke, Newton was well acquainted with the current theories of light, but dissatisfied with the explanations. The consensus was that ordinary light was homogenous producing the different colors by refraction when striking the retina. Moreover, it was thought that red and blue were the dominant colors produced by the strongest impact of ordinary light on the retina and that the other colors were mixtures of these. Unconvinced, Newton began his prism experiments that display his remarkable ingenuity and dexterity, as well as his attention to detail and gift for interpretation. What he concluded was that ordinary light when refracted through a prism disperses into a spectrum of colors. Thus rather than luminous light being homogeneous, it appeared to be composed of a mixture of rays of colors which the retina, as the prism, refracts into separate colors.

The conclusion that light was composed of a mixture of colors was reinforced when he redirected the divided rays onto another prism and found that they again became homogeneous. He also inferred that the rainbow, whose explanation had been a central topic of optics throughout history, was caused by droplets of water refracting light into a spectrum. In attempting to measure the dimensions of the different colors caused by thin transparent films he discovered that circular lenses produce rings of colors, now referred to as "Newton's rings." He retained the term "rays" to designate the individual colors, but unlike Hooke and Huygens, who accepted the wave theory of the propagation of light, Newton rejected that view for a corpuscular interpretation that he believed was supported by the sharp outlines of shadows instead of wavy lines, and because it agreed with his growing acceptance of a corpuscular-mechanistic conception of physical reality.

Thus the young Newton, after several millennia of previous research, attained a partially correct explanation of the nature of light. As Westfall states: "No other investigation of the 17th century better reveals the powers of experimental inquiry animated by a powerful imagination and controlled by rigorous logic." (p. 164) One will have to await the 19th century for the discovery that light is an electromagnetic radiation with wave properties and the 20th century for the further discovery that light consists of energy quanta called photons and thus behaves either as a wave or a particle depending upon the physical conditions, contributing to the name "wavicles."

Notwithstanding these accomplishments Newton again was faced with uncertainty regarding his future at Trinity College, the necessary election to a fellowship if he were to accede to a permanent position. However, even though his family qualifications had not changed, and his election depended upon the Master of the College and eight Senior Fellows who would have been expected to use the traditional unacademic criteria, he was elected. This might have been due to Humphrey Babington, one of the senior fellows, though because of the secrecy of the proceedings this is not known for certain. In any case, he was elected a Minor Fellow on 1 October, 1667 and then to a Major Fellow when he earned his Maser of Arts degree nine months later.

During his election to the Fellowship at Trinity, Isaac Barrow, the Lucasian Professor of Mathematics at Cambridge, had been asked to examine him on Euclid in connection with the Fellowship. As Newton had been studying Descartes' geometry rather than Euclid's, he thought he had not done well. Yet several years later when he showed Barrow his method for calculating infinite series, Barrow was so impressed that he sent John Collins, "a mathematical impresario who… functioned as a clearinghouse for information…to keep the growing mathematical community of England and Europe abreast of the latest developments," a copy of Newton's paper describing him as "'a fellow of our College & very young…but of an extraordinary genius & proficiency in these things.'" (p. 202)

This proved most opportune for Newton for he could not have been introduced to a person better suited to evaluate his genius and promote his reputation. Impressed with the method sent by Barrow, Collins proceeded to send Newton further intricate mathematical problems which he would solve within a month or so and return. Increasingly impressed, Collins sent copies to mathematicians in England, Scotland, and Europe, but when he urged Newton to have some published and encountered resistance, respecting Newton's aversion to publicity he discontinued the exchange.

Notwithstanding the termination, the correspondence had advanced Newton's reputation as a gifted mathematician which probably explains Barrow's decision to resign the Lucasian Professorship recommending Newton as his successor. Thus unlike Einstein whose academic ascension progressed more slowly, at the age of twenty-seven Newton attained the prestigious and lucrative chair of Lucasian Professor. As described by Westfall, the

> professorship...ranked behind the masterships of the great colleges and the two chairs in divinity...as the ripest plum of patronage in an institution much concerned with patronage. On 27 October, 1669, this plum fell into the lap of an obscure young fellow of peculiar habits, apparently without connections, in Trinity College—to wit, Isaac Newton. (p. 206)

But even such good fortune was not without its ordeal. In accepting the Lucasian Professorship Newton was required to "embrace the true religion of Christ with all [his] soul" and "take holy orders when the time prescribed by these statutes arrives [within seven years of incepting M. A.], or...resign from the college." (p. 179; brackets added) This meant having to uphold the Trinitarian doctrine and remain celibate. While the latter posed no problem because of his sexual orientation, with his usual dedication Newton spent a number of years studying early church history, particularly the Council of Nicaea's decision in the fourth century to adopt the Athanasian over the Arian Creed.

Concluding that the Athanasian doctrine of the consubstantiality of God the Father, Christ the Son, and the Holy Ghost was "a massive fraud," he adopted the Arian interpretation that Christ and the Holy spirit were created by God, but not of the same essence, and thus in good conscience could not swear to uphold the Athanasian Creed. Thus to avoid having to resign his chair at Trinity, a special dispensation was required that would exempt him from taking the oath of office. Again something undisclosed eventually intervened, perhaps in the personage of Isaac Barrow, so that on 27 April, 1665 the dispensation was official exempting the Lucasian professor "from taking holly orders unless 'he himself desires to...'" (p. 333) Not granted to Newton personally, it was to show "that His Majesty was willing 'to give all just encouragement to learned men who are & shall be elected to ye said Professor-ship....'" (p. 333)

In addition to his scholarly activities that were contributing to his growing reputation, another event testifies to his diversified talents. When performing his optical experiments he had discovered that reflecting telescopes had the disadvantage of producing chromatic aberration which resulted in poor focusing and thus decided to build a reflecting telescope to eliminate the distortion. Without any additional help he "cast and ground the mirror from an alloy of his own invention," as well as "the tube and the mount." (p. 233) Just six inches long, it still magnified objects nearly forty times in diameter with a clear image. (Westfall provides a drawing of it on p. 235) The invention is reminiscent of Galileo's construction of his telescope, though Newton created his without the assistance of a technician.

When learning of his improved invention and to commend his skill, he was proposed candidate to the Royal Society and duly elected on 11 January, 1672, the beginning of many honors to be bestowed on him in the coming years. In a letter to Henry Oldenburg, who was then Secretary of the Society thanking him for the nomination, Newton wrote that he would send to the Society the results of his prism experiments and new theory of colors which he declared was of more importance than the improvement of the reflecting telescope. Oldenburg's later acknowledgement was "filled with lavish praise" informing Newton that when read before the Royal Society it "'mett with both a singular attention and an uncommon applause'" and that the "Society had ordered that it be printed forthwith in the *Philosophical Transactions* if Newton would agree." (p. 239) Though he would regret it because of the later criticisms it would evoke from defenders of the wave theory, such as Hooke and Huygens, he agreed to its publication which was the first of many in the *Transactions*.

Nearly a decade after receiving his dispensation, a visit by Dr. Halley (of Halley's comet fame) in 1684, as recalled by Newton to Abraham DeMoivre, would serve as a catalyst in his development. Newton's theory of planetary motion being the purpose of the visit, Halley asked what he thought

> "the Curve would be that would be described by the Planets supposing the force of attraction towards the Sun to be reciprocal to the square of their distance from it. Sr Isaac replied immediately that it would be an Ellipse, the Doctor struck with joy and amazement asked how he knew it, why saith he I have calculated it whereupon Dr Halley asked for his calculation...." (p. 403)

The visit reigniting Newton's interest in planetary motions, he set out to construct a dynamic framework of celestial mechanics involving forces, another of his unique contributions since most astronomers, except for Kepler, were adverse to the concept of forces believing they were occult in nature. Then several months after his visit Newton sent Halley a short essay, *De motu corporum in gyrum* (*On the Motion of Bodies in Orbit*) in which he indicated that he not only had derived the elliptical orbit from the inverse square law, but now had inferred that an "elliptical orbit entails an inverse-square force to one focus...." (p. 404) The essay also included a discussion of the orbital periods of Jupiter's and Saturn's satellites and the moon, along with the cause of the tides and the motion of the comets. Following a series of revisions of the original nine page essay, he began expanding it into two books more than ten times larger with the title shortened to *De motu corporum* (*On the Motion of Bodies*) which, abbreviated, became the subtitle of Volume I of *Principia Mathematica*.

Nearing completion in the winter of 1686, the two enlarged books comprising Books I-III of the *Principia* when completed was submitted to the Royal Society for consideration for publication. As he described the process thirty years later: "'The Book of principles [the *Principia*] was writ in about 17 or 18 months, wereof about two months were taken up with journeys, & the MS was sent to ye R. S. [Royal Society] in spring 1686; & the shortness of the time in which I wrote it, makes me not ashamed of having committed some faults.'" (p. 444; brackets added) That such an outstanding work was written in such a short time further attests to Newton's genius.

Rather than sending the manuscript to the Royal Society himself, Newton entrusted it to Dr. Vincent, a member of the Society, who

> presented to the Society a manuscript treatise entitled *Philosophiae Naturalis Principia Mathematica*, and dedicated to the Society by Mr. Isaac Newton, wherein he gives a mathematical demonstration of the Copernican hypothesis as proposed by Kepler, and makes out all the phaenomena of the celestial motions by the only supposition of a gravitation towards the center of the sun decreasing as the squares of the distances therefrom reciprocally. (pp. 444-445)

Largely through the assistance and perseverance of Dr. Halley who oversaw the drafting of the book and preparation for publication, it was finally ready for publication on 5 July, 1687. Unlike Copernicus who did not acknowledge Rheticus' crucial role in persuading him to write the *De Revolutionibus* and arranging its printing, in the Preface to Vol. I, Newton graciously praised Halley (along with the Royal Society) for his assiduous assistance in seeing the book through its final draft and publication.

> In the publication of this work the most acute and universally learned *Mr. Edmund Halley* not only assisted me in correcting the errors of the press and preparing the geometrical figures, but it was through his solicitations that it came to be published; for when he had...obtained my demonstrations of the figure of the celestial orbits, he continually pressed me to communicate the same to the Royal Society, who afterwards, by their kind encouragement and entreaties, engaged me to think of publishing them.[54]

In addition to acting as midwife to Newton's *Principia,* Halley wrote an *Ode Dedicated to Newton* that ended with these moving lines:

> "*Then ye who now on heavenly nectar fare,*
> *Come celebrate with me in song the name*
> *Of Newton, to the muses dear; for he*

> *Unlocked the hidden treasuries of Truth:*
> *So richly through his mind had Phoebus cast*
> *The radiance of his own divinity.*
> *Nearer the gods no mortal may approach."*
> (p. xv)

Turning to the *Principia* itself, in a frequently quoted *Preface to the First Edition* Newton concisely describes its purpose:

> ...I offer this work as the mathematical principles of philosophy, for the whole burden of philosophy seems to consist in this—from the phenomena of motions to investigate the forces of nature, and then from these forces to demonstrate the other phenomena; and to this end the general propositions in the first and second Books are directed. In the third Book I give an example of this in the explication of the System of the World; for by the propositions mathematically demonstrated in the former Books, in the third I derive from the celestial phenomena the forces of gravity with which bodies tend to the sun and the several planets. Then from these forces, by other propositions which are also mathematical, I deduce the motions of the planets, the comets, the moon, and the sea. (pp. xvii–xviii)

The prophet that he was, he indicates he wishes that he could explain the structure of matter, but expects future generations by following his method will succeed in accomplishing, which indeed they have with the discovery of atomic elements, the molecular structures of substances, numerous subatomic particles and the strong and weak nuclear forces. Moreover, notice how modest Newton's claim is as compared to Bacon and Descartes.

> I wish we could derive the rest of the phenomena of Nature by the same kind of reasoning from mechanical principles, for I am induced by many reasons to suspect that they may all depend upon certain forces by which the particles of bodies, by some causes hitherto unknown, are either mutually impelled towards one another, and cohere in regular figures, or are repelled and recede from one another. These forces being unknown, philosophers have hitherto attempted the search of Nature in vain; but I hope the principles here laid down will afford some light either to this or some truer method of philosophy. (p. xviii)

Whether Newton adopted a rigorous formalistic presentation because it would appeal to mathematical physicists, ward off unqualified criticism, or was more congenial, in any case Book I begins with precise definitions of concepts we know as 'mass,' 'momentum,' 'inertia,' 'acceleration,' and 'centripetal force,' followed by a Scholium introducing his theory of absolute time, space, place, and motion. (Cf. pp. 6–7) Not only did he think that these absolute coordinates were justified by his calculations, he believed they were necessary to avoid the atheistic implications of Descartes' relativistic motions. As he confessed to the Reverend Bentley : "When I wrote my treatise about our Systeme, I had an eye upon such Principles as might work wth considering men for beliefe of a Deity" (Westfall, p. 441), considering absolute time, space, place, and motion more congruent with God's omnipotence.

These absolutes, inferred from direct experience, formed the invariant frames of the universe until the Michelson-Morley experiments and Einstein's theories of relativity introduced the relativization of spatial, temporal, and velocity measurements. According to Newton's explication of absolute motion and rest:

> Absolute motion is the translation of a body from one absolute place into another; and relative motion, the translation from one relative place into another. Thus in a ship under sail, the relative place of a body is that part of the ship which the body possesses.... But real, absolute rest is continuance of the body in that same part of that immovable space, in which the ship itself, its cavity, and all that it contains, is moved. (p. 7)

Curiously, his actual justification of their absolute natures was based on the shape of the surface of the water in a rotating bucket, believing that the concave ascent of the water up the sides of the bucket, as it was rotated, was evidence of centrifugal forces generated by the rotation in absolute space.

> The effects which distinguish absolute from relative motion are the forces of receding from the axis of circular motion.... If a vessel held at rest together with the water... is whirled about the contrary way...the surface of the water will at first be plain as before the vessel began to move, but after that, the vessel by gradually communicating its motion to the water, will make it begin sensibly to revolve...and ascend to the sides of the vessel, forming itself in a concave figure.... This ascent of the water shows its endeavor to recede from the axis of its motion; and the true and absolute circular motion of the water...becomes known and, may be meassured by this endeavor. (p. 10)

What this illustrates is not the naiveté of Newton but that abstractions derived from ordinary experiences are often misleading because they do not represent the more basic forces of the world. Sometimes these ordinary observations or common sense inferences are entirely false, while at other times they are viewed as approximations to the real state of affairs. The history of science is strewn with such discarded concepts as 'geocentrism,' 'crystalline spheres,' 'occult forces,' 'ether,' 'fixed species,' 'magnetic fluid,' and so forth. It is the fact that scientists have developed a method for detecting these deceptions and replacing them with testable, explanatory concepts and theories is its justification and crowning glory. Scientists are often criticized because they continually revise theories as new evidence becomes available, but this is one of the great merits of science in contrast to clinging to false beliefs.

The discussion in the Scholium is followed by his presentation of axioms or laws of Motion that systematize and culminate the astronomical laws discovered by Kepler and the gravitational laws experimentally proved by Galileo. It is these integraed laws that confirm Kepler's conception of a "mechanistic, clockwork universe." They also constituted the basis of much of the research conducted during the 18th and early 19th centuries until emended and partially replaced in the 20th century. Newton's formula $F = ma$ is still the accepted formula when dealing with macroscopic objects and slow velocities, but must be supplanted when considering atomic, subatomic, and cosmic dimensions illustrated by Einstein's equally famous formula $E = mc^2$.

Having presented his fundamental definitions and laws of motion, the next section of the *Principia*, BOOK I, THE MOTION OF BODIES, contains Newton's exact mathematical demonstrations, accompanied by numerous diagrams, Propositions, Theorems, and Lemmas describing the motion of bodies with his conception, especially, of gravitational forces. Like Galileo, given the primitive state of scientific explanations, he disavows speculating about the exact nature of the forces that produce the motions, but restricts himself to calculating their magnitudes.

> I here use the world *attraction* in general for any endeavor whatever, made by bodies to approach each other, whether that endeavor arise from the actions of the bodies themselves, as tending to each other or agitating each other by spirits emitted [as Kepler initially believed]; or whether it arises from the action of the ether or of the air, or of any medium...whether coreal or incorporeal, in any manner impelling bodies... towards each other. In the same general sense I use the word *impulse*, not defining in this treatise the...physical qualities of forces, but investigating the qualities and mathematical proportions of them.... (p. 192; brackets added)

Though his supporting diagrams are usually geometrical, occasionally he uses his theory of fluxions when discussing magnitudes approaching "vanishing limits." For instance, Aristotle's paradox as to how an object projected vertically can reverse its direction of motion while momentarily coming to rest at its peak, or how there can be instantaneous velocities which imply motion in durationless intervals, is explained by showing how the rate of a dependent variable, such as velocity or distance, can vanish as the independent variable, time or velocity, approaches the "limit" of zero: for example, the now familiar differential notation ds/dt. As his analysis of Aristotle's paradox states, it may be alleged that a body arriving at a certain place, and there stopping, has no ultimate velocity.... But the answer is easy; for by the ultimate is meant...that velocity with which the body arrives at its last place and with which the motion ceases. And in like manner, by the ultimate ratio of evanescent quantities is to be understood the ratio of the quantities not before they vanish nor afterwards, but with which they vanish. (pp. 38–39)

An excellent example of the way in which the mathematical formalism assists scientists in solving measurement problems, it indicates why it is so indispensable in physics, as Newton attests.

In mathematics we are to investigate the quantities of forces with their proportions consequent upon any conditions supposed; then, when we enter upon physics, we compare those proportions with the phenomena of Nature that we many know what conditions of those forces answer to the several kinds of attractive bodies. And this preparation being made, we argue more safely concerning the physical species, causes, and proportion of the forces. (p. 92)

This mathematical dependence was illustrated previously when he showed how the attractive gravitating force of the nearest massive object deflects planetary bodies from a natural rectilinear trajectory to an elliptical orbit permitting the deduction of the exact magnitudes of the distances, velocities, and forces to produce Kepler's first two laws, along with his third law that the squares of the times of their revolutions are proportional to the cubes of their mean distances from the sun. With Einstein's equation $E = mc^2$ we now realize why the sun, with 98% of the energy in our solar system, can be taken as the exclusive massive source of the gravitational forces causing the planetary motions in their respective orbits, simplifying the calculations.

In another major advance toward the end of BOOK I, he compares the radiation of light to the gravitational force (the converse of Kepler's early comparison) declaring that there is now evidence for rejecting the millennia old belief that light is transmitted instantaneously: "For it is now certain from the phenomena of Jupiter's satellites, confirmed by the observations of different astronomers, that light is propagated in succession, and requires about seven or eight minutes to travel from the sun to the earth." (p. 229) As further evidence of the relation of the two, he states that rays of light when passing near bodies "are bent or inflected round those bodies as if they were attracted to them...." (p. 230) While this discussion of BOOK I only covers a small fraction of Newton's investigations and explanations, it should be sufficient to illustrate the range, originality, and depth of his thinking. One begins to sense how the modern conception of the world was being formed beyond what others were able to achieve, despite his acknowledged limits of the explanations.

The SECOND BOOK of Volume I comprising nine sections deals with the effects of different media on the motion of bodies and the ratios involved, such as the resistance due to the compression and density of fluids, the effect of air on pendular motions, the properties of the particles in liquids affecting their fluidity, diffusion, and resistance, and the manner in which oscillating bodies propagate their motions in elastic media. These investigations are further evidence of his extensive curiosity and acuity. He ends BOOK II of Vol. I by refuting Descartes' theory of vortices which maintained "the periodic times of…the vortex to be as the square of the distances from the center of the motion" (p. 393) violating Kepler's third law that they varied with the $3/2^{th}$ power. He concludes that "the hypothesis of vortices is utterly irreconcilable with astronomical phenomena, and rather serves to perplex than explain the heavenly motions." (p. 396)

Despite his refutation, the theory of vortices consisting of a purely mechanical explanation based on the physical contact of bodies, rather than Newton's so-called 'occult force' of gravitation acting at a distance, had strong adherents among natural philosophers, such as Huygens, Perrault, and Bernoulli. Visiting England to learn of Newton's cosmological system, Voltaire wrote: "A Frenchman who arrives in London finds a great alteration in philosophy, as in other things. He left the world full, he finds it empty [absolute space]. At Paris you see the universe composed of vortices and subtle matter, in London we see nothing of the kind."[55]

Turning now to Volume II containing BOOK III of the *Principia*, with the subtitle *THE SYSTEM OF THE WORLD* (the basis of Hooke's charge of plagiarism because this was same title he had used for one of his early works), Newton had intended it to be a nonmathematical popularization of his natural philosophy, but later recast it in the same rigorous mathematical form to avoid controversy by those who were incapable of understanding it. As he writes in the introductory paragraph:

> In the preceding books I have laid down the principles of philosophy; principles not philosophical but mathematical: such, namely, as we may build our reasonings upon in philosophical inquiries.... It remains that, from the same principles, I now demonstrate the frame of the System of the World. Upon this subject I had, indeed, composed the third Book in a popular method, that it might be read by many; but afterwards, considering that such as had not sufficiently entered into the principles could not easily discern the strength of the consequences, nor lay aside the prejudices...which they had been many years accustomed, therefore, to prevent the disputes which might be raised upon such accounts, I chose to reduce the substance of this Book into the form of Propositions (in the mathematical way), which should be read by those only who had first made themselves masters of the principles...in the preceding Books....[56]

It is significant that while he wrote of laying "down the principles of philosophy," he insisted that they were "principles not philosophical but mathematical," illustrating the difference between his conception of scientific principles and those of Bacon and Descartes, despite referring to them as "principles of philosophy."

This is followed by the four "RULES OF REASONING IN PHILOSOPHY" that will underlie and guide his scientific inquiries. Rule *I* affirms Occam's razor that explanatory principles should not be multiplied beyond necessity: "*We are to admit no more causes of natural things than such as are both true and sufficient to explain their appearances.*" Rule II proclaims the uniformity of nature: "*Therefore to the same natural effects we must, as far as possible, assign the same causes.*" Rule III justifies accepting the primary properties of objects as more real than their secondary or sensory qualities: "*The qualities of bodies, which admit neither intensification nor remission of degrees, and which are found to belong to all bodies within the reach of our experiments, are to be esteemed the universal qualities of all bodies whatsoever.*" Rule IV maintains that the empirical foundations of science require the exclusion of any nonscientific hypotheses: "*In experimental philosophy we are to look upon propositions inferred by general induction from phenomena as accurately or very nearly true, notwithstanding any contrary hypotheses that may be imagined, till such time as other phenomena occur, by which they may either be made more accurate, or liable to exceptions.*" (pp. 398–400; italics in original) These explicit rules of reasoning are crucial for they have guided scientific inquiry ever since!

Rule III addresses the question raised by the ancient Greek philosophers as to what is the true nature of physical reality. Leucippus and Democritus proposed that it consisted of atoms, discrete insensible particles shorn of *sensory qualities* but defined as hard, indivisible, shaped, and mobile *physical qualities*, interacting according to mechanistic principles. Aristotle, in contrast, rejected atomism for Empedocles' theory of the four elements, fire, air, earth, and water, along with denying mechanistic explanations in favor for his four causes, material, formal, efficient, and final, while Bacon suggested spirit as constituting the inner nature of objects. It was Aristotle's more accessible theory that prevailed in Western thought from the reintroduction of his philosophy by Thomas Aquinas in the 13th century to the 17th. Newton, however, favored the mechansitic-corpuscular view of the atomists in his interpretation of natural phenomena influenced by Robert Boyle's explanation of his gas laws as caused by the kinetic motion of the insensible particles composing the gases.

The construction of the microscope in the latter part of the 17th century disclosing the existence of microstructures and particles also contributed to the revival of the atomic theory. But the dispute continued into the early 20th century with such leading scientists as Ernst Mach still denying the existence of atoms despite the discovery of subatomic particles such as electrons and protons, along with alpha, beta, and gamma rays. However, most scientists became convinced of the existence of microscopic particles after Einstein's explanation of Brownian motion and Ernest Rutherford's and Niels Bohr's initially successful model of the atom composed of the recently discovered electrons, protons, and neutrons.

The following statement contains Newton's own justification of his belief in minute insensible particles and the corpuscular theory.

> We no other way know the extension of bodies than by our senses, nor do these reach it in all bodies; but because we perceive extension in all that are sensible, therefore we ascribe it universally to all others also. That abundance of bodies are hard, we learn by experience; and because the hardness of the whole arises from the hardness of the parts, we therefore justly infer the hardness of the undivided particles not only of the bodies we feel but of all others. That all bodies are impenetrable, we gather not from reason, but from sensation.... The extension, hardness, mobility, impenetrability and inertia of the whole, result from the extension, hardness, mobility, impenetrability, and inertia of the parts; and hence we conclude the least particles of all bodies to be... extended, and hard and movable and impenetrable, and endowed with their proper inertia. And this is the foundation of all philosophy. (p. 399)

As mentioned previously in our discussion of Galileo, since John Locke and Newton were close friends this could be the source of Locke's well-known distinction between primary and secondary qualities.

Rule IV presents Newton's distinction between "feigned" hypotheses, such as Bacon's spirit and Descartes' vortices that were conjectural, as opposed to hypotheses supported by empirical and experimental evidence and inductive laws. When he added a GENERAL SCHOLIUM to BOOK III of the second edition of the *Principia* published in 1713, he was even more emphatic in rejecting all "metaphysical" hypotheses which he described as not being "deduced from the phenomena," accepting the existence of gravity based on the astronomical evidence, though objecting to formulating speculative hypotheses as to its exact nature:

> ... hitherto I have not been able to discover the cause of those properties of gravity from phenomena, and I frame no hypotheses; for whatever is not deduced from the phenomena is to be called an hypothesis, and hypotheses, whether metaphysical or physical, whether of occult qualities or mechanical, have no place in experimental philosophy. In this philosophy particuular propositions are inferred from the phenomena, and afterwards rendered general by induction. (p. 547)

However, he seemed to have revised his objection to hypotheses in a letter to Henry Oldenburg, Secretary to the Royal Society, quoted by Cajori, in which he condoned the use of hypotheses if limited to explaining the properties of objects *after* they have been discovered by experiment, but not before.

> "For the best and safest method of philosophizing seems to be, first diligently to investigate the properties of things and establish them by experiment, and then to seek hypotheses to explain them. For hypotheses ought to be fitted merely to explain the properties of things and not…to predetermine them except in so far as they can be an aid to experiments. If anyone offers conjectures about the truth of things from the mere possibility of hypotheses, I do not see how anything…can be determined in any science; for it is always possible to contrive hypotheses one after another, which are found rich in new tribulations." (p. 673)

Here he wisely does not reject all hypotheses, but only those arrived at by conjecture, rather than deduced from experimental evidence.

Because pseudosciences such as alchemy, astrology, and occultism were still prevalent one would admire Newton for his sage advice except, unfortunately, that he was not always consistent in following it. Newton not only had an extensive library of alchemy, but secretly pursued the pseudoscience for many years. As he states,

> alchemy…is not of that kind wch tendeth to vanity & deceit but rather to profit & to edification inducing first ye knowledge of God & secondly ye way to find out true medicines in ye creatures….so y^t ye scope is to glorify God in his wonderful works, to teach man how to live well, & to be charitably affected helping o^r neighbors. (Westfall, p. 298)

One wonders how he could have accepted alchemy as a true science, considering its flawed methodology and hypotheses, and acceded to these implausible claims.

Even more flagrantly, whereas previously he had stated that he was not going to speculate whether the "forces" or "impulses" causing the motion of bodies consist of "spirits," "ether" or "air," surprisingly he now resembles Bacon in introducing "a certain subtle spirit," which he claims "pervades and lies hid in all gross bodies," as the cause of all interactions, whether magnetic, electrical, optical, or neurological. But later, fortunately, he will replace this "subtle spirit" with an "Æthereal Medium."

> And now we might add something concerning a certain most subtle spirit which pervades and lies hid in all gross bodies; by the force and action of which spirit the particles of bodies attract one another at near distances, and cohere if contiguous; and electric bodies operate to greater distances, as well repelling as attracting…neighboring corpuscles; and light is emitted, reflected, refracted, inflected, and heats bodies; and all sensation is excited, and the members of animal bodies move at the command of

the will, namely, by the vibrations of this spirit, mutually propagated along the solid filaments of the nerves from the outward organs of sense to the brain, and from the brain into the muscles. But these are things that cannot be explained in a few words, nor are we furnished with that sufficiency of experiments that is required to an accurate determination and demonstration of the laws by which this electric and elastic spirit operates. (p. 547)

Although contradicting his previously rejection of unempirical hypotheses, if I am not being too partial this could be seen as a prescient statement anticipating in prospect, if not in content, the present search for a Grand Unified Theory (GUT). We perhaps can excuse him at this early time for thinking that this force might be a spiritual one considering his attraction to alchemy, although his suggestion that it could be "electric and elastic" contains at least a hint of the "electromagnetic" or "electrodynamic field" suggested by some contemporary physicists in their search for the unifying force in GUT. Moreover, it could be seen as foresight that Newton attributes nerve discharges to the same "vibrating spirit" occurring "along the solid filaments of the nerves, from the outward organs of sense to the brain," that produces the other phenomena he describes.

He concludes Vol. II of the *Principia* with a GENERAL SCHOLIUM in which a final account is given of how the orbits of the planets with their satellites follow from Kepler's laws confirmed by the astronomical observations of John Flamsteed, the Royal Astronomer. He shows that Kepler's elliptical orbits require the sun being at their center, the final confirmation of Copernicus's heliocentric hypothesis, while also proving that the planet's center of gravity is apart from, but close to, the sun. A final explanation is given of the motion of the comets and cause of the earth's tides.

Thus ends the *Principia*, universally acknowledged as the greatest scientific treatise ever written, whose theoretical framework for understanding the world guided scientific research until the early 20th century, but whose exact prescription for the correct methodology of science is still valid. At the time, even the French mathematician of considerable repute, Marquis de l'Hôpital, could wonder whether it had been written by an ordinary mortal. When showed the *Principia* by Dr. John Arbuthnot,

he cried out with admiration Good god what a fund of knowledge there is in that book? he then asked the D^r every particular about Sr I. even to the colour of his hair said does he eat & drink & sleep. is he like other men? & was surprised when the D^r told him he conversed cheerfully with his friends assumed nothing & put himself upon a level with all mankind. (Returning to Westfall, p. 473, as are the directly following quotations).

Before turning to his final work, the *Opticks,* there are three episodes in his life that reveal more of his character. The first involves King James II, a Catholic, who came to the throne in 1685 determined to make Catholicism the established religion of the English Anglicans. Believing this could best be accomplished by permitting Catholics to obtain positions of authority at the universities which was then prevented by their having to take "the oath of supremacy, in effect an oath to uphold the established Anglican religion" (p. 474), he decided to remove this obstacle by using the traditional "letters of mandate" to confer higher degrees on Catholics exempting them from taking the oath.

To test his authority the King proposed a Benedictine Monk, Alban Francis, to the degree of Masters of Art at Cambridge University. When the Vice Chancellor John Peachell decided to resist, Newton drafted a supporting letter urging "an honest Courage" which would "save ye University." (p. 475) When he received the dissenting letter the King summoned Peachell and a faculty delegation to which Newton and Humphrey Babington were elected, along with eight others, to appear before the Court of Ecclesiastical Commission headed by Lord Jeffreys. When the King in a compromise proposed that Father Francis be awarded the degree on condition that this not be considered a precedent, Newton strongly objected persuading the delegation that this would be a dishonorable capitulation and dangerous precedent.

Unfortunately, when the group met before the Commission Peachell was so intimidated by Lord Jeffreys that he was unable to present a strong defense of the delegation's position and, in retribution, the King removed him from all his offices at Cambridge. With Peachell forced to resign from the University, the opposition was left to the other members of the delegation with Newton again forcefully advocating that they should not give in, drafting five letters preceding the final written refusal, including in one draft the statement that a "mixture of Papist & Protestants in ye same University can neither subsist happily nor long together" (p. 479) which, however, was not included.

The delegation duly met with Lord Jeffreys and the Commission without knowing if they faced the same fate as poor Peachell, but it was the Commission this time who relented with Jeffreys' conceding but warning that in the future they must obey His Majesty's commands. However, the threat proved futile because James II was disposed eighteen months later by William of Orange and fled to France where he was unable to influence English affairs.

Newton's fortitude and wise council did not go unnoticed with his colleagues immediately showing their admiration. When it came time for two delegates to be elected to represent the University at the convention to ratify the Glorious Revolution, Newton was chosen as one. In addition, he was directed by acts of

Parliament to be one of the regular Commissioners to oversee the collection "in Cambridge of aids voted to the government" (p. 480), a lucrative appointment indicative of his enhanced standing. This changed Newton's life not only by increasing his income, but also by permitting his moving to London for a year when the convention was reconvened as Parliament. It was due to his move to London that he met Christian Huygens and John Locke with whom he formed a close friendship. This must have had a strong influence on Locke's philosophy whose epistemological views in *As Essay Concerning Human Understanding* were directed opposed to those of Descartes, but congruent with Newton's conception of scientific knowledge.

This elevated position, along with the acclaim brought by the publication of the *Principia*, produced a period of "manic euphoria" from 1687-1693 followed by a "blackyear" beginning in the autumn of 1693 owing to his breakup with Fatio de Duillier, a young Swiss mathematician whom he had met in 1687. This apparently was a repeat of an earlier experience he had in 1677-1678 when the intimate relationship he had formed at Trinity College with his chamber fellow, John Wickins, came to an end. The second occurred when Duillier, who was twenty years his junior, wrote to Newton saying "'I could wish Sir to live all my life, or the greatest part of it, with you, if possible'" (p. 533), but not receiving an affirmative reply he formed another attachment. This break plunged Newton into a nervous depression, as intense as his previous state of euphoria, that lasted for a year and a half until his recovery.

During this time he decided to write a book based on the results of his optical experiments performed thirty years earlier. When the final draft was completed in 1694 and he showed it to his friend David Gregory, the latter declared that it "would rival the *Principia*." When the Royal Society offered to publish the work, Newton retracted it owing to his controversy with Hooke over the latter's charge of plagiarism related to the subtitle of Volume II of the Principia. But when Hooke died in 1703 and Newton was elected President of the Royal Society he offered the *Opticks* to the Society for publication, with Halley again chosen to make the decision which was a foregone conclusion.

Though the book did not rival the *Principia* in originality since it contained little that was new regarding his former optical experiments, because it made less mathematical demands on the reader, contained much fewer complex diagrams, and was originally written in English rather than Latin, it was more accessible to a broader public and thus more widely read. Still, the kinds of questions it raised were so original and extensive that they were the basis of much of the experimental research throughout the 18th century. Among these queries were the investigation of the reflection, refraction, inflection, and colours of light; the visual system including its nervous discharges to the brain; the nature of heat, radiation, and magnetism; electrical attraction and repulsion; gravitational forces; chemical reactions; and the microstructure of substances.[57] Some of these topics had been addressed in the *Principia*, but others were new and seemed to embrace the entire range of current scientific inquiry.

Foreshadowing Rutherford's use of alpha particles to probe the interior of gold foil detecting the existence of the proton, Newton radiated what he believed to be light corpuscles on various substances to disclose their interior structures, along with attempting to measure the dimensions of the corpuscles themselves. As I. Bernard Cohen states in his excellent book, *Franklin and Newton*, contrasting the *Principia* and *Optics*:

> Not primarily in the *Principia*, then, but in the *Opticks* could the eighteenth century experimentalists find Newton's methods for studying the proper ties or behavior of bodies that are due to their special composition. Hence we need not be surprised to find that in the age of Newton—which the eighteenth century certainly was!—the experimental natural philosophers should be drawn to the *Opticks* rather than to the *Principia*.[58]

Since much of the content of the *Opticks* was previously mentioned in the discussion of his earlier experiments, only a brief recounting will be given here. The book contains a Preface, an extensive Introduction, an Analytical Table of Contents, followed by three Books which are further divided into Parts with sub-headings such as Definitions, Axioms, Propositions, and Observations, followed by 31 Queries.

In Book I he states that he is not going to try to explain the properties of light by introducing hypotheses, but instead describe and prove the existence of these properties by experiments. He redescribes his prismatic experiments in which he discovered that ordinary light consists of "Primary, Homogeneal and Simple Rays." These Rays, when reflected from the surface of objects, pass through the pupil, are refracted by the crystalline lens and the humors, converging "in the bottom of the Eye, and there to paint the Picture of the Object upon that skin (called the *Tunica Retina*) with which the bottom of the eye is covered.... And these Pictures, propagated by Motion along the Fibres of the Optick Nerves into the Brain, are the cause of Vision." (p. 15) He distinguishes between the "Power and Disposition" of the physical rays to transmit their motion to the Sensorium and the "Sensations [caused there by] these Motions" which are experienced "under the Forms of Colours." (pp. 124–125; brackets added). He also discusses the causes of near and far sightedness, jaundice, and again chromatic aberration in refracting telescopes.

Book II describes his numerous experiments reflecting and refracting light in prisms and convex lenses, as well as in air and water, discovering that when the lenses are very thin the light forms a series of colored rings, now called "Newton rings," as mentioned previously. He carefully measured the width of the layers of air separating the thin plates of "plano-convex glass," along with the distances between the successsive colored rings and their diameters, finding that when he "measured the Diameters of the first six Rings at the most lucid part of their Orbits, and squaring them," that their Squares were "in the arithmetical Progression of the odd Numbers, 1, 3, 5, 7, 9," (p. 200), surprisingly repeating Galileo's odd number law of gravitational acceleration.

Along with these ingenious measurements, he tried to calculate the dimensions and distances of the discrete particles within the air and water, as well as the forces causing the pressure, heat, and dispersion of the gases, perhaps the first known effort to analyze the microstructure of substances by radiational probing. As he declared: "there are many Reflections made by the internal parts of Bodies, which…would not happen if the parts of those Bodies were continued without any such Intersices between them…." (p. 249). He presents a table "wherein the thickness of Air, Water, and Glass, at which each Colour is most intense and specified, is expressed in parts of an Inch divided into ten hundred thousand equal parts." (pp. 232–233) Inferring the diameters of the interior parts or corpuscles from the different colors of the light reflected is, probably, one of the first efforts to find experimental evidence of the corpuscular or atomic theory antedating Einstein's explanation of Brownian motion, again exhibiting Newton's extraordinary ingenuity.

Notwithstanding these ingenious experiments and complex calculations, he was aware that they were not sufficient to reveal the inner structure of substances: "what is really their inward Frame is not yet known to us." (p. 269) But rather than suggesting improved probing experiments, he claimed that with more powerful microscopes one should be able to see the actual particles, which is possible today with tunneling microscopes. This reference to more acute microscopic observations could have been the source of Locke's term "microscopical vision."

Though Newton rejected the wave theory of light, his experimental discovery that "Light reflected by thin Plates of Air and Glass" and then directed through a prism "appear *waved* with many Successions of Light and Darkness made by alternate Fits of easy Reflection and easy Transmission, the Prism severing and distinguishing the Waves of which the white Light was composed" (p. 281; italics added), indicated that under certain conditions light could manifest wave properties. Had he not been so committed to the corpuscular theory, he might have anticipated the present dualistic conception of light displaying either particulate or wave properties depending upon the experimental arrangements. Though this was only a brief sweep of the contents of Book II, it should have been sufficient to illustrate the scope of Newton's inquiries.

Book III containing the 31 Queries also attests to the exceptional range, originality, and depth of his reflections. As I. Bernard Cohen affirms: "To the eighteenth-century reader, as to us, these queries reveal the mind of Newton in its innermost thoughts just as the reading of the book of Nature revealed to Newton the mind of the creating God." (Cohen, op. cit., p. 177) Usually starting with "Are not...," "Do not...," or "Is not...," they read less as inquiries than as reflective conclusions that would provide much of the research incentives of the 18th century. Today when we are constantly confronted by new discoveries in physics, astrophysics, chemistry, biology, neurophysiology, and electronics, it is easy to overlook how unusual such a barrage of challenging questions were at that time. Unlike the Aristotelians who thought all his inquiries were final or Galileo who had expressed his skepticism regarding the possibility of solving such questions, for Newton the prospects appeared endless.

In addition to his former remarkable discoveries, he seems to have come close to anticipated Planck's explanation of blackbody radiation when he observed that the reflection of light produced by heating black bodies is due to the change in the intensity of the internal vibrations of the heated particles. Having suggested in Query 6 that "Black bodies conceive heat more easily from Light than those of other colours do" (p. 339), he asks in Query 8: "Do not all [black] Bodies, when heated beyond a certain degree, emit Light and shine; and is not this Emission perform'd by the vibrating motions of their Parts?" (p. 340: substituting 'black' for 'fix'd') What a remarkable question! The closeness to Planck's explanation is even more apparent when we recall that Newton had adopted the corpuscular theory of light over the wave theory, which would have allowed him to explain the emission of light producing the "shine" by the radiation of light corpuscles activating the vibration of the internal particles.

Investigating vision, he distinguished between the right and left optic nerves, concluding that single vision is due to the uniting of the two nerves in what we know as the optic chiasm, though he mistakenly thought that the stimuli from the right and left visual fields were conveyed to the corresponding hemispheres of the brain, rather than the opposite sides. He made extensive investigations of Island Crystal attributing the polarization of the rays of light when refracted in Island Crystal to the different "Sides of the Rays to the Planes of perpendicular Refraction." (p. 359) Distinguishing transverse waves produced when stones are dropped in water from the waves caused by percussion instruments, he asserted that the planes in Island Crystal put the refracted rays into definite orientations perpendicular to one another, so that if refracted through a second crystal the rays will either be transmitted or deflected depending upon the alignment of the planes of the crystal. In addition, he attempted to explain the alignment by the arrangement of the particles in the Island Crystal.

Though incorrect because today polarization is attributed to the wave properties of light, Newton's perspicacious adoption of particles and forces enabled him to foresee many future causes of natural phenomena in terms of a substructure of particles and forces or charges, as indicated in Query 31:

> Have not the small Particles of Bodies certain Powers, Virtues, or Forces, by which they act at a distance, not only upon the Rays of Light for reflecting, refracting, and inflecting them, but also upon one another for producing a great part of the Phœnomena of Nature? For it's well known, that Bodies act one upon another by the Attractions of Gravity, Magnetism, and Electricity; and these Instances shew the Tenor and Course of Nature, and make it not improbable but that there may be more attractive Powers than these. For Nature is very consonant and conformable to herself. (pp. 375–376)

It should not be forgotten that Newton was almost alone in extolling the use of "Powers, Virtues, or Forces" in explaining natural phenomena (ironically due to his alchemical inquiries), since most continental natural philosophers rejected them as occult, especially if they were thought to act at a distance as Newton claimed. Probably he was at his most prophetic when he declared that whatever these powers or attractive forces are "it is the Business of experimental Philosophy to find them out." (p. 394) Surely the advance of science has been a confirmation of this claim.

As at the close of Vol. II of the *Principia*, where he speculated whether there might be a "subtle spirit" within all material bodies explaining their motions, near the end of the *Opticks* he raises the same question but rejects a "subtle spirit" for an elastic "Æthereal Medium." Thus he asks in Query 18: "is not this Medium exceedingly more rare and subtile than the Air, and exceedingly more elastick and active? And doeth it not readily pervade all Bodies? And is it not (by its elastick force) expanded through all the Heavens?" (p. 349) Another precursor of a later theory, it anticipated the ether theory of the 19th and early 20th centuries that, as a rarified substance, was believed to be necessary for the transmission of light and electromagnetic waves and therefore pervaded the universe, until Einstein proved that it was as unnecessary. Moreover, replacing his former "subtle spirit" by a quasi-material substance indicates Newton's increasing belief that natural processes could be explained entirely by principles derived from induction and experiments, rather than by spiritual hypotheses, a progression that also had been true of Kepler.

> To tell us that every Species of Things is endow'd with an occult specifick Quality by which it acts and produces manifest Effects, is to tell us nothing: But to derive two or three general Principles of Motion from Phænomena, and afterwards to tell us how the Properties and Actions of all corporeal Things follow from those manifest Principles, would be a very great step in Philosophy, though the Causes of those Principles *were not yet discover'd*: And therefore I scruple not to propose the Principles of Motion above-mention'd, they being of very general Extent, and *leave their Causes to be found out*. (pp. 401–402; Italics added)

As Bacon and Descartes, he believed that scientific explanation depends upon discovering the right "principles," but he differed in how these were acquired. Yet he added that at the time of creation they had to be supplemented by "the Counsel of an intelligent Agent," a forerunner of the argument from "intelligent design."

> Now by the help of these Principles, all material Things seem to have been composed of the hard and solid Particles…variously associated in the first Creation by the Counsel of an intelligent Agent. For it became him who created them to set them in order. And if he did so, it's unphilosophical to seek for any other Origin of the World, or to pretend that it might arise out of a Chaos by the mere Laws of Nature [as Anaximander maintained]; though being once form'd, it may continue by those Laws for many Ages…. (p. 402; brackets added)

This deistic conception of God, that once He had created the universe it could be left to operate according to the created laws, became popular among many scientists at the time, and apparently was believed by Einstein who often said that in trying to understand the laws of physics we are trying to understand the mind of God. Also, in his earlier book, *A Brief History of Time* the famous cosmologist, Steven Hawkings wrote that if we could answer the "question of why it is that we and the universe exist," then "it would be the ultimate triumph of human reason—for then we would know the mind of God." But in his recent book, *The Grand Design*, written with Leonard Mlodinow, he describes the cosmological theory of "multiuniverses" which would contain a vast number of alternate universes whose laws of nature would be very different from ours, and therefore produce entirely different physical environments. Thus if this model of the universe proves true, there would be no need to evoke the Mind of God as the first cause because science could provide the necessary and sufficient conditions for the origin of this world.[59]

There are a few (pseudo) scientists, such as William A. Dembski and Michael Behe, along with attorney Phillip Johnson, who believe that the creation of species or the complexity of cellular structures presupposes an intelligent Agent or designer as their cause. Yet in his book, *Finding Darwin's God*,[60] cell biologist

Kenneth R. Miller provides *extensive scientific evidence* to *disprove* the claims made by these anti-Darwinians to support their arguments, despite then declaring that "[t]he God of the Bible, even the God of Genesis, is a Deity fully consistent with what we know of the scientific reality of the modern World" (p. 258) which, as we have just seen, certainly does not agree with the claims of leading physicists and cosmologists.

Convinced that there is not enough empirical evidence to decide cosmological issues as there is to support Darwin's theory, Miller uses human experience to describe whatever attributes, capabilities, and motives are required to explain how and why God brought about the creation of the universe. As a result, his reasoning appears naively anthropomorphic and fabricated, as if God were just an exalted human craftsman, even though he admits that ultimately we cannot understand God's actions. It is difficult not to conclude that if he had not been coaxed since childhood to believe in the existence of God it is very unlikely that, as a scientist, he would have found it reasonable to introduce a divine creator to supplement his scientific explanations.

This judgment is reinforced by his startling admission that a "key doctrine in my own faith is that Jesus was born of a virgin, even though it makes no scientific sense—there is the matter of Jesus' Y-chromosome to account for. But that is the point. Miracles, by definition, do not have to make scientific sense." (p. 239). So miracles can be combined with scientific explanations? This is a striking example to what extremes religionists must go to justify irrational beliefs, accepting miracles as reconciling the blatant divergences between the two worldviews which is hardly credible. Better to give up "what makes no sense," rather than to continue to believe what is sheer nonsense.

Returning to our recounting of Newton's life, having completed the *Opticks* in 1994 he delayed having it published until 1704 following the death of Hooke the previous year. After that he decided to forsake the scholarly life he had enjoyed at Trinity College for thirty-five years to live in London permanently. With the loss of his university salary and privileges and even with his recently increased income, this was not sufficient to support living in London, so he sought a more lucrative position. After several disappointments, with the help of his friend Charles Montague he obtained the position of Warden of the Mint on 19 March, 1696. Owing to his skillful administration, three and a half years later he was promoted to Master of the Mint, a position he occupied until his death. This resolved his financial problems because, along with his annual salary, he received the considerable sum of "a set profit on every pound weight troy that was coined" (returning to Westfall, p. 604). As a consequence, the following year he was ready to resign his fellowship at Trinity and the Lucasian Chair at Cambridge.

His motivation for leaving the academic world and moving to London having been to lead a more active political and social life, this was initially fulfilled by his position at the Mint entailing membership in the House of Commons. At the urging of Montague, now Lord Halifax, he also stood for parliament and was elected for one year but defeated the following year. It was owing to this political endeavor, not for his academic achievements, that he received his knighthood from Queen Anne. Succeeding her husband William to the throne following his death, she conferred knighthood on Newton during a visit to Cambridge to enhance his political prospect in the next election. But even the title of "Sir Isaac" did not improve his standing as he came in last among four candidates with few votes, thus thwarting any political ambitions.

However, when he became Warden of the Mint in 1696 there occurred an incident more congruent with Newton's talents when Johann Bernoulli, a leading European mathematician, published a challenge problem in Wallis' *Acta eruditorum* in June with the condition that it had to be solved within six months. Having received only a single response by December from Leibniz, who replied that he had solved the problem but suggested that the challenge be redistributed and the time limit extended to Easter of the following year, Bernoulli agreed.

Adding a second problem to the previous challenge, he sent both to the *Philosophical Transactions* of the Royal Society and also to the *Journal des scavans* to ensure that they would be seen by both British and European mathematicians. He also sent copies to Newton and Wallis. Shortly after receiving his copy Newton solved the problems, returning them anonymously to Bernoulli who had no difficulty identifying the author due to the method he used to solve them. Only Leibniz, the Marquis de l'Hôpital, and Newton submitted solutions. Obviously Newton had not lost his mathematical acuity despite his older age and radical change of life.

Among his accumulated honors, he was offered membership in the Académie Royale des Sciences of Louis XIV which carried a large pension, but for whatever reason he declined the invitation. The Masterships of St. Catherine's Hospital and of Trinity College also were offered to him but they too were rejected, the latter because he would have had to take the Orders. But when Hooke died in 1703 and the presidency of the Royal Society became vacant, he decided to seek the office. Though nominated, his being elected was not a foregone conclusion as one might have supposed since he barely received the maximum votes, apparently because his attendance during the years Hooke was president was minimal because of their quarrel.

Yet when elected he served with his usual distinction such that the attendance, membership, quality of lectures, and financial support improved considerably. For example, Hooke's death having revoked their privilege of meeting at Gresham College where he had resided, Newton, with the aid of the Secretary Hans Sloane, persuaded the members to purchase the former house of Dr. Edward Browne on Crane Court off Fleet Street for its permanent residence. The house was bought in 1710 and was extensively renovated according to the design and supervision of Christopher Wren, a member of the Royal Society. "In less than six years, the society had fully paid for its new home and stood free of debt," so for "the first time in the fifty years since its establishment, the Royal Society had its own home." (Westfall, p. 677) That no longer is the residence of the Society which now is located in the attractive row of white buildings on 6-9 Carlton House, near the prestigious Athenaeum Club.

When Newton was elected president of the Royal Society on 30 November, 1703 following Hooke's death the previous March, he was sixty-one years old and would live for another quarter of a century less one year. During most of that time he remained generally in good physical and mental health until about five years before his death when his condition began to deteriorate, though he continued in his positions as Master of the Mint and President of the Royal Society until the end. During those later years his previous devotion to natural philosophy and mathematics was completely replaced by his lifelong interest in biblical studies. As Westfall states: "Newton set out at an early age to purge Christianity of irrationality, mystery, and superstition, and he never turned from that Path" (p. 826), though he also never succeeded. Yet despite his belief in God as the creator of the universe (understandable for the time) and fascination with alchemy, like Kepler he finally adopted a rigorous scientific method that served as a model for future scientists.

This rigorous scientific method not only led to an increasingly reduced credibility of religious beliefs, it also offered the means of alleviating many of the disasters, such as famines, plagues, draughts, poverty, and diseases, that formerly induced them. Whereas throughout the past religious beliefs were the primary means of alleviating the suffering caused by such catastrophes by claiming they had some divine purpose, despite their devastating effects conflicting with the conception of an all loving and all powerful God, the new scientific method made it possible to eliminate or moderate their consequences. The tremendous progress made in medicine that has greatly reduced infant and maternal mortality, along with childhood diseases, considerably raising the longevity of human beings in developed societies, is a striking example.

As asserted previously, throughout history (and for religionists today) most eventualities were believed to be the result of *intentional causes* willed by supernatural beings, as in the Christian saying "thanks be to God" or the Muslim expression

"God Willing," explaining the reliance on animal or human sacrifices, religious rites and prayer, along with absolutions, penance, and indulgences as a way of gaining favor with the gods. That the evangelists Pat Robertson and Jerry Falwell attributed several national disasters to God's punishment for supporting abortion and gay rights is evidence of the continuation of this primitive way of thinking.

It was discovering their natural causes making it possible to eliminate or at least ameliorate them, thereby gaining more control over our lives, *that has reduced our dependence on a Supreme Being and the reason for believing in one.* The extraordinary advances in science and technology during the 20th century, especially, enhanced this change. Does anyone really believe that the traumatic situations confronting us, such as the worldwide economic crisis, terrorist attacks, overpopulation, energy shortages, declining natural resources, climate change, and natural disasters, will be solved by prayer, rather than by intelligent means aided by scientific inquiry and technology? It is thanks to the originators of modern science just described that made these improvements possible.

In striking contrast to the self-aggrandizement of Bacon and Descartes and despite his incomparable achievements and honors, the following statement by Newton made toward the end of his life reveals his basic humility and character (which is true of the greatest scientists): "I don't know what I may seem to the world, but, as to myself, I seem to have been only like a boy playing on the seashore, and diverting myself in now and then finding a smoother pebble or a prettier shell than ordinary, whilst the great ocean of truth lay undiscovered before me." (Westfall, p. 863) Though his theoretical framework that guided scientific research for two centuries is now seen to apply only to limited conditions, his meticulously formulated methodology still stands as a beacon for scientific progress.

Newton died after a short illness on 20 March, 1727 at age eighty-five and is interred in a prominent location in the nave of London's Westminster Abbey. His monument bears the fitting inscription: "Let Mortals rejoice that there has existed such and so great an Ornament to the Human Race." So ends the saga of those who had the intelligence, courage, and perseverance to devise the methodology that made it possible for humanity to rise above superstition, mythology, ignorance, and religious dogma, along with redressing those conditions that have brought so much suffering to mankind in the past. Such was their gift of enlightenment.

Epilogue

For those readers who are only interested in the origins and development of early modern science and how it has undermined the supernatural foundation of religion, they might want to discontinue reading further. However, lest the reader be left with the impression that the Newtonian worldview represents the last stage in the development of science, in this Epilogue I will describe, as succinctly as possible, the discoveries that led to a radical revision, mainly in the twentieth century, of the Newtonian cosmological system with the realization that, despite its remarkable achievements, it applies only within certain limited conditions, and therefore represents just a beginning stage in the advances of science.

The primary developments consist of the repudiation of Newton's completely deterministic universe by discoveries in quantum mechanics and Einstein's rejection of Newton's absolute space and time and reinterpretation of his theory of gravity. The rejection of determinism began with Max Planck's introduction of the quanta of energy in 1900 followed in 1905 by Einstein's denial of Newton's absolute space and time in his special theory of relativity, and then his replacement of Newton's gravitational force by a four-dimensional space-time field in his general theory of relativity in 1915. The latter was confirmed in 1919 by the telescopic observations of Arthur Eddington on the island of Principe off the coast of West Africa and in Brazile.

The radical transformations of quantum mechanics began with Planck's introduction of energy quanta to explain blackbody radiation that led to Einstein's explanation of the photoelectric effect by discrete quanta of light later called 'photons.' It had long been known that when a blackbody, such as a poker, is heated it changes color from red to violet to glowing white with the rise in temperature. Believing that the changes in color were due to the intensified radiation of the vibrating particles within the poker caused by the increase in heat-energy (as Newton had proposed), Planck was astonished to find that the *interaction* occurred in discrete amounts or 'quanta,' not continuously as expected, introducing 'quanta of energy' to account for the results. Although this agreed with the experimental data, he never was reconciled to his explanation in terms of discrete units of energy.

It was not until 1905, when Einstein published his paper on the photoelectric effect during his *annus mirabilis* while working in the Patent Office in Bern, Switzerland, that the quantization of energy and light began to be accepted. Produced by shining ultraviolet light on the surface of certain metals, attempts to explain the effect as due to the emission of electrons proved unsuccessful in terms of the continuous wave structure of ultraviolet light. Knowing of Planck's explanation of blackbody radiation incorporating discrete units of energy rather than waves, Einstein turned to his theory and formula of quanta of energy to provide an explanation.

He found that by replacing the light *waves* interacting with the metallic oscillators with light *quanta* whose energy was defined by Planck's formula, $E = h\nu$, correlating the energy E with the frequency ν times Planck's constant h, gave a correct explanation. The specific amount of energy required to activate the oscillators to reflect the *frequency* of a particular color, red requiring less and violet more, was described in terms of the magnitude of *specific* light quanta, while the *intensity* of the illumination was explained as due to the *number* of the quanta striking the metallic surface. Though Planck had restricted the quantization to the energy interaction, Einstein proposed that the light itself was radiated as quanta of energy later named 'photons,' an acronym of 'electrons.' Thus the wave-particle duality entered physics, one of the most perplexing paradoxes of quantum mechanics.

While physicists were puzzling over the fact that light waves mysteriously behave like particles, a young French Prince named Louis de Broglie submitted a doctoral dissertation at the Sorbonne in 1924 suggesting that since waves can behave as particles, perhaps particles can behave as waves, which he designated as "matter waves." Then, using Einstein's formula $E = mc^2$ showing the energy E of a particle equals its mass m times the velocity of light squared c^2 and Planck's formula $E = h\nu$, he derived his own formula $\lambda = h/mv$ where the wave length λ equals Planck's constant h divided by the mass m times velocity v (momentum), "the celebrated *de Broglie matter wave relation.*"[61] In the same year Bohr and two of his colleagues, H. A. Kramers and John Slater, suggested that since the waves in diffraction experiments had a probable distribution, they should be called "probability waves." Two years after de Broglie had submitted his dissertation proposing matter waves Clinton Davisson and his assistant, Lester Germer, verified it experimentally. (Cf. p. 74) Thus the two most confounding aspects of quantum mechanics, the wave-particle duality and probability, entered the fray.

In 1921 Bohr had founded the Institute for Theoretical Physics in Copenhagen, sponsored by the Carlsberg Brewery, which would play such a central role in the development of quantum mechanics as nearly every important contributor visited the Institute at some time during their career, with many receiving their early training there. So it was that the prodigy Werner Heisenberg began his studies at Bohr's Institute also in 1924. The following year he, along with Max Born and Pascual Jordan, published a paper on matrix mechanics in the *Zeitschrift für Physik* that was followed in 1925 by a more comprehensive paper on the mathematics of matrix mechanics.

Based entirely on measurements, this mathematical formalism produced probable predictions of the trajectories of radiated particles *without indicating what physical processes were involved.* Precluding the visualization of subatomic structures and processes, it called in question the existence of such atomic structures as Bohr's solar model of the atom. It also had the mathematical oddity that multiplying numbers was no longer commutative: while in ordinary mathematics $3 \times 5 = 5 \times 3$, this is not true of matrix mathematics where 3×5 does not equal 5×3. The peculiarity of its mathematics and exclusion of visual models to depict subatomic processes made matrix mechanics unpopular with more traditionally oriented physicists, especially Einstein. Nonetheless, in 1925 the "*infant terrible*," Wolfgang Pauli, would attempt to elucidate the structure of the hydrogen atom with matrix mechanics.

Later Heisenberg also became convinced that the difficulties encountered in quantum mechanics could be eliminated by discarding theoretical models such as Bohr's solar model of the atom and restrict explanations entirely to the mathematics based solely on the scattering of particles produced in collision experiments of high-energy physics called the "Scattering Matrix" or S Matrix". (p. 81)

At the time Heisenberg was creating matrix mechanics, Erwin Schrödinger, who preferred a visual model of the atom, was developing an alternative mathematical interpretation of quantum mechanics influenced by de Broglie's matter waves, called "wave mechanics," that initially proved more congenial to most physicists. He devised an equation containing a mathematical quantity known as a wavefunction symbolized by the Greek letter Ψ (*psi*) which, when squared

$|\Psi|^2$, gave the probability of locating an electron somewhere in space. (Cf. p. 89) Unlike the presumed precise location of previous particles during their radiation, the probable path of the "wavefunction" or "packet" is spread over the entire space until measured, which then "reduces it to a singularity." This is also described as a "superposition" of all the possibilities before measurement which, when measured, collapses into a single state.

The difference between Schrödinger's wave mechanics and Heisenberg's S matrix mechanics is that while the latter only predicts the probabilities of the quantum mechanical interaction, excluding the physical causes, the former describes these probabilities in terms of pictorial wave trains produced by the initial conditions which collapse into definite magnitudes when measured. But, surprisingly, as different as were their conceptions of the physical transformations, the mathematics of Schrödinger's wave mechanics was found to be equivalent to Heisenberg's matrix mechanics.

Then Heisenberg, focusing on the measurements themselves, introduced a principle that would *disallow the preexistence* of physical properties, whether of waves or particles, *independently of their being measured*, the most radical assault on Newton's worldview. Traditionally, as a prerequisite of any scientific inquiry it was presupposed that the independent particles with "inherent physical properties" existed locally. But influenced by the probability measurements in quantum mechanics, Heisenberg introduced his "principle of uncertainty," claiming that at the *dimensional level* of quantum mechanics the classical presuppositions of causal determinism, locality, and scientific realism were no longer valid!

He found that in measuring certain conjugate attributes, such as position and momentum or energy and time, the more exact the measurement of one, the less certain the measurement of the other: $\Delta p \times \Delta q > h/2\pi$ and $\Delta e \times \Delta t > h/2\pi$. (Cf., pp. 95, 99 respectively). What this indicates is that the product of the probabilities of the measured conjugates must be greater than Planck's constant $h/2\pi$ or $(\hbar)$ which has the incredibly tiny value in units of kilograms, meters, and seconds of 6.63×10^{-34}. Being so infinitesimal it does not apply on the scale of macroscopic or even atomic measurements, but on subatomic dimensions presumably accounting for its indeterminacies.

Initially this indeterminacy was attributed to the disturbance caused by the interaction of the detector with the minute particle being measured: for example, the exact measurement of a particle's position affecting its momentum and the measurement of its momentum affecting its position, so that the more precise the measurement of one, the less precise it is of the other. Later, however, it was attributed either to the mathematical formalism involving probability waves or to the fact that such properties do not having an independent existent until measured.

In response to these challenges to his earlier model of the atom and to the hidden variable view, Bohr discarded his earlier pictorial solar model for his influential "theory of complementarity." Conceding that what is known about the quantum world is limited to experimental measurements that provide only probable or indeterminate knowledge, he affirmed that we must discard the assumption that physical reality at that level of inquiry consists of *preexisting* physical elements and their properties and accept only the evidence provided by the mathematical formalism. He offered the example of a blind man exploring the world with a cane to illustrate how the observer and the observed are entwined through the act of detection.

As stated by Manjit Kumar in the American Edition of *Quantum* published in 2010:

> Bohr thought that quantum mechanics was a complete fundamental theory of nature, and he built his philosophical worldview on top of it. It led him to declare: 'There is no quantum world. There is only an abstract quantum mechanical description. It is wrong to think that the task of physics is to find out how nature is. Physics concerns what we can say about nature.'[62]

As investigations of subatomic particles disclose either wave or particle properties under different experimental conditions, they must be seen as "complementary" aspects dependent on the measurements, since such contradictory features cannot coexist as independent, inherent properties of a quantum entity!

This was his answer to both Schrödinger and Heisenberg, as well as Einstein, that became for a time the unquestioned interpretation of quantum mechanics. It challenged Einstein to show, by various thought experiments, that the uncertainties in quantum mechanics are an indication of its *incompleteness* which could be explained in relation to a deeper level of physical reality, as in the past, "the hidden variable theory." In contrast, Bohr claimed that rather than quantum mechanical explanations being emended by further discoveries, the only conception we have of this reality is indicated in the experimental results as represented in the mathematical formalism. And since there has been no disconfirming evidence, it must be complete.

A final development seemed to have confirmed Bohr's rejection of hidden variables and locality. In 1935 Einstein, along with Nathan Rosen and Boris Podolsky, published a paper in the *Physical Review* entitled, "Can Quantum Mechanical Description of Physical Reality be Considered Complete?" Although only four pages long, it became instantly famous because it prescribed an experiment intended to refute Heisenberg's uncertainty principles and Bohr's Copenhagen interpretation. It's stated thesis is: "'*If, without in any way disturbing a system, we can predict the certainty (i.e. with probability equal to unity) the value of a physical quantity, then there exists an element of physical reality corresponding to this physical quantity.*'"[63]

Described as simply as possible, the EPR article, as it is called, presents a thought experiment involving the interaction of two particles, A and B. Heisenberg's uncertainty principle precludes a particle having both an exact position and momentum *at the same time* because they cannot be measured simultaneously, only separately, their existence thus dependent on their being measured. In opposition, the authors of the EPR experiment argue that having *interacted* the two particles were correlated so that one could *infer* either the position or the momentum of particle B from measuring either the position or the momentum of particle A, however distantly separated. Because they could choose either of the two properties to measure, they inferred that implied they both must preexist in particle B prior to being measured, contrary to the thesis of both Heisenberg and Bohr. Not allowing this, they concluded that quantum mechanics must be incomplete.

Bohr's rebuttal proved more difficult than he had anticipated. Eventually he settled on refuting the central premises of the EPR paper, that the measurements on B occurred "without in any way disturbing the system...." Bohr insisted that the initial "interaction" of the two particles were so *correlated* that the inferred properties of particle B were actually dependent on the measurements of particle A. Thus the *whole context* must be considered when assigning exact properties which, therefore, do not exist prior to the measurements. That the two properties exist in a kind of interdependence was later called "entanglement" by Schrödinger and disparaged by Einstein as "spooky actions at a distance," but "spooky" or not it played a crucial role in the later discussions.

To further test the two theories, in 1951 David Bohm proposed a simplified EPR experiment based on George Uhlenbeck and Samuel Goudsmit's discovery in 1925 that in addition to its other properties, an electron has a spin with only two possible states, spin up and spin down. The electron's original state has zero spin, but when it disintegrates it forms two electrons with opposite spin directions. Analogous to the earlier EPR experiment, the original entanglement allowed a prediction of the second particle's orientation without measuring it after determining that of the first, again supporting the EPR thesis of the independence of the properties from measurement. Bohr in contrast argued that since it is the measurement that caused the collapse, it is not until A is measured that "its entangled partner B acquires the opposite spin...even if it is on the other side of the universe" (p. 344), affirming his nonlocal Copenhagen interpretation.

Then in 1964 a young Irish physicist from Belfast, John Stewart Bell, decided to see if there was still a way of confirming the local hidden variable theory. Based on the electron's spin, Bell devised a theorem by which one could compare the predictions of any local hidden variable theory with those of quantum mechanics by measuring the correlations of pairs of spinning electrons by a detector at various settings. As spin theory does not imply that the first measurement continues to determine the second particle's orientation by extending a cause over vast distances of space, it should be possible to distinguish between the two explanations by their correlations.

Modifying Bohm's experiment, Bell measured the maximum number of spin correlations of two entangled electrons permitted by the hidden variables theory called "Bell's inequality," and finding that the quantum mechanical correlations *exceeded* those allowed by his inequality concluded this disproved the hidden variables interpretation. Yet other explanations being possible, Bell urged further experiments to settle the issue. In 1969 John Clauser took on the challenge with the assistance of Stuart Freedman, switching to polarized photons with the orientation that if one is polarized upward, the other is polarized downward, while their combined polarization is zero.

In 1975 they tested Bell's inequality by heating calcium atoms until they acquired enough energy for an electron to jump from a ground state to a higher energy level. Returning to the ground state, the electron did so in two stages emitting a pair of entangled photons, one green and the other blue. The photons were deflected in opposite directions with detectors set to measure simultaneously their polarizations. Initially both detectors were set at 22.5 degrees, then separately aligned by 67.5 degrees and the two sets of measurements compared. After 200 hours of measurements, Clauser and Freedman found "that the level of photon correlations violated Bell's inequality…in favour of Bohr's nonlocal Copenhagen interpretation of quantum mechanics with its 'spooky action at a distance,' and against the local reality position backed by Einstein." (p. 348)

Still not convinced, a French graduate student, Alain Aspect, after studying Bell's experiments and consulting him in Geneva, decided as his doctoral thesis to make a further attempt to test the alternatives. Between 1981 and 1982 Aspect performed three experiments intending to verify Bell's inequality in *support* of hidden variables, using the latest laser and computer technologies. Though replicating Clauser's experiment he enhanced the results by increasing the number of entangled pairs of photons measured, but was forced to conclude that the evidence indicated "'the strongest violation of Bell's inequalities ever achieved, and excellent agreement with quantum mechanics.'"[64] Yet Bell remained "convinced that 'quantum theory is only a temporary expedient' that would eventually be replaced by a better theory," despite conceding at the time of his death in 1990 "that experiments had shown that 'Einstein's world view is not tenable.'" (p. 350)

Given the misgivings regarding the correct interpretation—though not the vallidity of the mathematical formalism—Richard Feynman made the famous declaration in 1965, ten years after Einstein's death: "I think I can safely say that nobody understands quantum mechanics…. 'Do not keep asking yourself, if you can possibly avoid it, "but how it can be like that?" 'Nobody knows how it can be like that.'"[65] Einstein, however, persisted in believing that since the past advances of science had relied on the discovery of deeper levels of physical reality to explain previously inexplicable phenomena, this still must be true.

Even more important was his motivation, in devising his two theories of relativity, to preserve the *independence of natural laws*, such as the invariant velocity of light, from the relative positions of observers. This was why he could never reconcile himself to Bohr's interpretation of quantum mechanics that its mathematic formalism, rather than representing the independently existing properties of quantum particles, only encompassed the formalism itself, that is, "to what could be said about them." In contrast, Einstein wanted to discover the laws God used in creating the world.

But despite no conclusive demonstration of the uncertainties in quantum mechanics, the indeterminacies led to the acceptance of Bohr's Copenhagen interpretation as the most plausible interpretation available. For his continued objection to this interpretation, Einstein was accused of being too old to accept the new theories and even of being senile. But later discoveries in subatomic physics began to call in question Bohr's interpretation, as when Murray Gell-Mann and Yuval Ne'man introduced a new model for analyzing hadrons, called "the eightfold way," consisting of quarks with such whimsically named *measurable* properties as 'up,' 'down' 'strangeness,' 'color' and 'flavor.'[66]

One weakness in Bohr's nonrealist, nonlocality interpretation was that it precluded any possible explanation of cosmological questions as to the origin of the universe, although the WMAP and COBE satellites and the Hubble space telescope were providing unprecedented evidence to explain the origin and structure of the universe. As Michio Kaku states in his intriguing book, *PARALLEL WORLDS*:

> With the flood of new data we are receiving today, with new tools such as space satellites which can scan the heavens, with new gravity wave detectors, and with city-size atom smashers nearing completion physicists feel that we are entering what may be the golden age of cosmology on this quest to understand our origins and the fate of the universe.[67]

The theory of parallel universes was constructed on Hugh Everett III's bizarre theory that rather than there being no existing physical correlate to quantum mechanical measurements, whenever a measurement is made every possible outcome is created in another world, the "many worlds interpretation" or what Hawking calls the "multiuniverse." According to this theory, ours is only a tiny fragment of one of these alternative universes.

Other developments included the Big Bang theory of the origin of the known universe introduced by George Lemaître in 1927 based on the spectroscopic discovery of the red shift in the light spectra from outer space. Interpreted as due to the recession of galaxies and thus of an expansionary or inflationary universe, it was shown by Roger Penrose and Howard Robinson to have been predicted by Einstein's general theory of relativity.[68] The detection by Arno Penzias and Robert Wilson in 1964 of the background microwave radiation left over from the Big Bang offered additional confirmation, while the evidence for black holes and dark matter and energy suggest that we know more about the universe than allowed in Bohr's "no quantum world" interpretation of quantum mechanics. String theory has offered an entirely new theoretical explanation of the universe, supporting Einstein's later suggestion that rather than the hidden variables theory completing quantum mechanics, a completely new theory would be required.

Analogous to the Pythagorean account of the universe, string theory consists of a cosmic harmony composed of vibrating strings, as described by Brian Greene.

> What appear to be different elementary particles are actually different "notes" on a fundamental string. The universe—being composed of an enormous number of these vibrating strings—is akin to a cosmic symphony.... If we can work out precisely the allowed resonant vibrational patterns of fundamental strings – the "notes," so to speak...we should be able to explain the observed properties of the elementary particles. For the first time, therefore, string theory sets up a framework for explaining the properties of the particles observed in nature.[69]

If true (a big if) physicists have come a long way from Bohr's position that we can know nothing about the properties of quantum particles to proposing that we can even *explain* them. Disappointingly, however, no evidence has yet been found to support string theory since its introduction in 1980, despite concerted research. But there were other developments that suggest that Bohr's interpretation was too limiting. In the 19th century Michael Faraday experimentally proved the causal conjunction of electrical and magnetic phenomena introducing the theory of electromagnetism, while James Clerk Maxwell created the mathematical equations describing the structure and propagation of the electromagnetic fields of force as a structure of spacetime.

Then, in addition to gravitational and electromagnetic forces two additional forces were discovered in the 1930's: first, a "strong force" acting over a short distance that binds the hadrons (the nuclear protons and neutrons), counteracting the electromagnetic repulsive force, and accounting for the creation of new particles when probed. A second "weak force" controls the interactions between the hadrons and leptons (the latter consisting of electrons, muons, neutrinos, etc.) in the radioactive decay of the nucleus. Quantum electrodynamics, QED, explains the effects of these forces, along with the organization of the atomic elements in the periodic table.

Thus while particles may not possess unconditional intrinsic properties in the classical sense, they do exist with their usual attributes *within* their causal interactive contexts even though this occurrence is conditional or contextual—as is true of ordinary perceptual and macroscopic qualities—yet are sufficiently objective to be the basis of explanatory investigations, as I argued in my book *Contextual Realism* cited previously. According to Gerard Piel:

> That has to be the case. The quantum world would otherwise be out of reach of experience. The instruments that excite quantum events make them perceptible in sought for mechanical, electric or electromagnetic effects. Surfacing, as they do, in such effects, quantum events are to be perceived at all times everywhere.[70]

Furthermore, there has been a concerted effort to attain a "grand unified theory," named "GUT." The weak force having been unified theoretically and experimentally with the electromagnetic force into an "electroweak force," there is now an attempt to unite this electroweak force with the strong force and gravitational force into a single, uniform mathematical framework. Thus the search for unification has led to the tentative conclusion that "Ordinary matter reduces…to four different particles, two each of two kinds, quark and lepton, that have so far proved irreducible.… All the diversity of nature in common experience arises from…their interaction through the four forces." (p. 117)

If all goes as expected at CERN, The European Organization for Nuclear Research, when the acceleration of the colliding particles reaches that of the first seconds of the Big Bang the "particle physicists expect to find the conditions projected in GUT theory that unite the three quantized forces [apparently excluding gravitational] in one primordial force." (p. 18; brackets added) They also hope to discover the Higgs Boson, or "God particle," that is postulated to have endowed the original massless particles with their mass. "Failing the ultimate union, so far beyond the reach of experiment, the "standard model" of particle physics and General Relativity separately comprehend all that experiment and observation can say about the physical world." (p. 119) (An interesting use of Bohr's expression 'say' rather than 'know.')

But just as the unexpected invention of the first telescopes and spectroscopes, followed by Hubble's series of advanced telescopes and space satellites, disclosed entirely new stellar bodies, such as billions of galactic universes beyond our Milky Way, while nuclear reactors and particle accelerators led to the discovery of subatomic particles, who can foretell the kinds of radical revisions in our worldview and living conditions that will be brought about by new experimental, theoretical, and technological advances? For example, George Charpak won a Nobel Prize in 1992 for developing "the multiwire proportional chamber, a particle detector that used computers to collect data 1,000 times faster than previous devices," that revolutionized particle physics.[71]

But since these physics and cosmological investigations pertain to only the 4 percent of the matter in the universe that we confront, the other 96 percent made up of 23 percent dark matter and 73 percent dark energy about which little is known, there surely will be many more unexpected theoretical developments. I essentially agree with the statement made by Steven Weinberg, in section 5 of his review of Hawking's latest book, *THE GRAND DESIGN*, mentioned in the previous chapter.

> Questions about the nature of reality have puzzled scientists and philosophers for millennia. Like most people, I think that there is something real out there, entirely independent of us and our models…. But this is because I can't help believing in an objective reality [as was true of Einstein], not because I have good arguments for it. (Brackets added)

Instead of searching for a final theory, perhaps the general view of "contextual realism" suggested previously is our best guess as to the status of the physical world in relation to both our perceptual experience and investigative inquiries of nature. In any case, despite the occasional impasses or disappointments in our research and the unlikelihood of attaining a final theory, at least the window of exploration will remain open for as long as there is human curiosity and a yearning to know.

End Notes

1. Ferris, Timothy, *The Science of Liberty* (New York: HarperCollins Publishers, 2010).
2. Singer, Charles, *A Short History of Scientific Ideas to 1900* (New York: Oxford University Press, 1960), pp. 62–63. The immediately following citations and all further references to Singer are to this book.
3. Clark, Gordon H., *Selections from Hellenistic Philosophy* (New York: Appleton-Century-Crofts, 1940), p. 219.
4. Cf. Schlagel, Richard H., *From Myth to Modern Mind: A Study of the Origins and Growth of Scientific Thought*, Vol. I, *Theogony through Ptolemy* (New York: Peter Lang Publishing, Inc., 1995, p. 419. The immediately following reference are to this work until otherwise indicated.
5. Plotinus, *The Enneads*, trans. by Stephen MacKenna, sec. ed., revised by B. S. Page (London: Faber and Faber Limited, 1956), V, 2:1; p. 380. The reference cites the standard manuscript references plus the page number in MacKenna. Similar page number citations will identify the references as from MacKenna.
6. Cf. Schlagel, Richard H., *The Vanquished Gods: Science, Religion, and the Nature of Belief* (Amherst, New York: Prometheus Books, 2001), Chapter 6, "Critique of Religious Experience."
7. Persinger, Michael A., *Neurophysiological Basis of God Beliefs* (New York: Praeger, 1987), p. x. The immediately following quotation is also to this work.
8. Slater, E. and A. Beard, *The Schizophrenic-like Psychoses of Epilepsy*. I. Psychiatric Aspects, *British Journal of Psychiatry*, 109 (1963), pp. 95–150; brackets added.

9. Nicholson, R. A., *Translation of Eastern Poetry and Prose* (Cambridge: Cambridge University Press, 1922), pp. 3–40.
10. Durant, Will, *The Story of Civilization*, Part IV, *The Age of Faith* (New York: Simon and Schuster, 1950), p. 163.
11. Saleh, Faiza, "Saudis Look Beyond Oil to New Economy in Desert," *The Washington Post*, July 17, 2008, A1.
12. Quoted from Artz, Frederick B., *The Mind of the Middle Ages* (New York: Alfred A. Knopf, 1962), p. 82.
13. Quoted from Taylor, H. O., *The Medieval Mind*, Vol. I, fourth ed. (Cambridge: Harvard University Press, 1962), p. 73.
14. Cf. Vantrease, Brenda Rickman, *The Illuminator* (New York: St. Martin's Griffin, 2005). Though a novel, it is based on extensive research involving historical figures that graphically and realistically portrays life as it was lived during the 14th century.
15. Arnaldez, R. and I. Massignon, "Arabic Science," trans. by A. J. Pomerans in René Taton, ed., *History of Science: Ancient and Medieval Science* (New York: Basic Books Inc., 1963), pp. 385–386. For a more comprehensive history of Arabic science, see Jim Al-Khalili, *Pathfinders: The golden Age of Arabic Science* (London, Allen Lane, 2010).
16. Quoted from Majid Fakhry's 1998 Internet article, "Greek Philosophy: Impact on Islamic Philosophy," www.muslimphilosophy.com, pp. 1–6. This brief article provides an excellent summary of the extensive Arabic translations of Greek philosophical texts, along with their impact on Arabic mathematical and scientific research. Unless otherwise indicated, the following quotations are to this article without indicating the page numbers.
17. This resumé is based on my *Seeking the Truth: How Science Has Prevailed Over The Supernatural Worldview* (Amherst, New York: Prometheus Books, 2010), pp. 203–209.
18. Crombie, A. C., *Robert Grosseteste and the Origins of Experimental Science 1100–1700* (Oxford: At the Clarendon Press), p. 12. All following references citing Crombie are to this work.
19. Aristotle, *Physics*, Richard McKeon, editor, *The Basic Works of Aristotle* (New York: Random House, 1941), p. 240. All the following references to Aristotle are to this standard compilation of his works.
20. Clavelin, Maurice, *The Natural Philosophy of Galileo*, trans. by A. J. Pomerans (Cambridge: The MIT press, 1974), p. 65. The following references to Clavelin are to this excellent study.
21. Newton, Isaac, *Principia Mathematica [Mathematical Principles of Natural Philosophy]*, Vol. I, *The Motion of Bodies*, trans. by Andrew Mott in 1729 and revised by Florian Cajori (Berkeley and Los Angeles: Univ. of California Press, 1934), p. 13.
22. Kuhn, Thomas, *The Copernican Revolution* (Cambridge: Harvard University Press, 1957), p. 122.
23. White, Andrew D., *A History of the Warfare of Science with Theology in Christendom* (New York: Appleton Press, 1896), p. 126. Quoted from Kuhn, op. cit., p. 191.
24. Copernicus, Nicolaus, *On the Revolutions of Heavenly Spheres*, trans. by Charles Glenn Wallis (Amherst, New York: Prometheus Books, 1995), p. 5.
25. This description of the five geometric solids are from my *From Myth to Modern Mind: A Study of the Origins and Growth of Scientific Thought*, Vol. I, *Theogony through Ptolemy*, op. cit., pp. 288–290.

26. Kepler, Johannes, *Mysterium Cosmographicum*, trans. by A. M. Duncan (New York: Abaris Books, 1981), p. 63. Unless otherwise indicated, the immediately following references are to this work.
27. Field, J. V., *Kepler's Geometrical Cosmology* (Chicago: University of Chicago Press, 1988), p. 38.
28. Koestler, A., *The Sleepwalkers* (New York: The Universal Library, 1963), p. 258.
29. Cf. Gilbert, William, *De Magnete*, trans. by P. Floury Mottelay (New York: Dover Publications., 1948), chs. III and IV. The present Dover edition is an unabridged and unaltered reprinting of the P. Fleury Mottelay translation published in 1893. All the following references to Gilbert are to this work unless otherwise indicated.
30. Letter to David Fabricius, "Johannes Kepler," *Gesammelte Werke*, Vol. XIV, p. 409. Quoted from Koestler, op. cit., p. 330.
31. Kepler, Johannes, Book 3, ch.V, *Harmonice Mundi*, R. M. Hutchins, ed., *Great Books of the Western World*, Vol. 16 (Chicago: The Univ. of Chicago Press), p. 1020.
32. Kepler, Johannes, *Astronomia Nova*, "Introduction," *Gesammelte Werke*. Quoted from Koestler, op. cit., p. 337.
33. Einstein, Albert and Leopold Infeld, *The Evolution of Physics* (New York: Simon and Schuster, 1951), pp. 156.
34. Drake, Stillman, *Galileo at Work: His Scientific Biography* (Chicago: The University of Chicago Press, 1978), p. 41. The following references are to this excellent work unless otherwise indicated.
35. Galileo Galilee, *Siderus Nuncius or The Sidereal Messenger*, trans. by Albert Van Helden (Chicago: University of Chicago Press,1989), p. 37. The immediately following references are to this work until otherwise indicated. I capitalize the names of the planets when the authors do.
36. Quoted from Stillman Drake, op. cit., p.199. All the immediately following references are to this work until otherwise indicated.
37. Cf. Schlagel, Richard H., *Contextual Realism: A Meta-physical Framework for Modern Science* (New York: Paragon House Publishers, 1986).
38. Galileo Galilei, *The Controversy on the Comets of 1618*, trans. by Stillman Drake and C. D. O'Malley (Philadelphia: University of Pennsylvania Press, 1960), p 53. This volume also contains Galileo's *The Essayer (Il Saggiatore)*, thus the immediately following references are to this work until otherwise indicated.
39. Galileo Galilei, *Dialogue Concerning the Two Chief World Systems -- Ptolemaic and Copernican*, trans. by Stillman Drake (Berkeley: University of California Press, 1962), p. 37.
40. Galileo Galilei, *Dialogues Concerning Two New Sciences*, trans. from the Italian and Latin by Henry Crew and Alfonso de Salvo (New York: McGraw-Hill Book Company, 1933). Until otherwise indicated, the following quotations are to this work.
41. Koyré, Alexander, *Galileo Studies*, translated by John Mepham (New Jersey: Humanities Press), 1978. For the quotation and the sources for Mersenne, see p. 126, f.n. 177; for Descartes, p. 107 and source p. 126, f.n. 176; and for Koyré, p. 107.
42. Bacon, Francis, *A Selection of his Works*, ed. by Sidney Warhaft (New York: The Odyssey Press, 1965), p. 1. All citations are to this work unless otherwise indicated.
43. Bacon, Francis, *The New Organon*, ed. by Fulton H. Anderson (New York: The Bobbs-Merrill Company, 1960). The immediately following citations are to this work.

44. Cf. Schlagel, Richard H., "The Waning of the Light: The Eclipse of Philosophy," *The Review of Metaphysics*, Vol. LVII, No. 1, Issue No. 225, September, 2003.
45. For an excellent recent account of Descartes' life and work see Geneviève Rodis-Lewis, *Descartes: His Life and Thought*, trans. by Jane Marie Todd (Ithaca: Cornell University Press, 1995). The immediately following biographical account and references are to this work until otherwise indicated.
46. Haldane, Elizabeth S. and G. R. T. Ross, *The Philosophical Works of Descartes*, Vol. I (Cambridge: At the University Press, 1967), PREFATORY NOTE. Unless or until otherwise indicated, the following references are to this work.
47. Cf. Rodis-Lewis, Geneviève, op. cit., p. 64.
48. This discussion is based on my article, *The Waning of the Light: The Eclipse of Philosophy*, cited previously, f.n. 44.
49. Westfall, Richard S., *Never at Rest: A Biography of Isaac Newton* (Cambridge: University Press, 1980), p. 59. The immediately following discussion and references are to this excellent work until otherwise indicated.
50. Koestler, A., *The Sleepwalkers*, op. cit., p. 406.
51. Small, Robert, *An Account of the Astronomical Discoveries of Kepler* (Madison, Wisconsin: The University of Wisconsin Press, l963), p. 305.
52. "'Newton to Hooke, 5 February 1676; Corres. 1, 416.'" Westfall, op. cit., p. 274.
53. Westfall, Richard S., op. cit., p. 403. Again the following references in the text are to this work until otherwise indicated.
54. Newton, Isaac, *Principia Mathematica*, Vol. I, *The Motion of Bodies* and Vol. II, *The System of the World*, Motte's trans., revised by Florian Cajori (Berkeley: University of California Press, 1962), p. XVIII. Until otherwise indicated the citations are to this work.
55. Voltaire, *Eléments de la Philosophie de Newton*, 1783. Quoted by Cajori, *Principia Mathematica*,Vol. II (Berkeley: University of Californian Press, 1962), p. 632; brackets added.
56. Newton, Isaac, Vol. II, *Principia Mathematica*, Book III, *System of the World*, op. cit., p. 397. The following references will be to this work unless otherwise indicated.
57. Newton, Isaac, OPTICKS: *A Treatise of the Reflections, Refractions, Inflections, & Colours of Light*. Based on the fourth ed. London, 1730 (New York: Dover Publications, 1952), p. 349ff. Unless otherwise indicated the following references are to this work.
58. Cohen, I. Bernard, *Franklin and Newton* (Cambridge: Harvard Univ. Press, 1966), p. 120. Further references to this work will be cited by the author's name and page.
59. For the first quotation see *A Brief History of Time* (New York: Bantam Books, 1988), p. 175. His recent book, THE GRAND DESIGN, was published by Bantam Books in September, 2010). For an excellent review of the book see Steven Weinberg's article, "The Universe We Still Don't Know," in The New York Review of Books, February 10, 2011. It also can be read on the Internet.
60. Cf. Miller, Kenneth R., *Finding Darwin's God* (New York: Harper Perennial, 2002), pp. 20, 60, 108.
61. Guillemin, Victor, *The Story of Quantum Mechanics* (New York: Charles Scribner's Sons, 1968), p. 73. The Following citations are to this work until otherwise indicated.
62. Kumar, Manjit's, *Quantum* (New York: W. W. Norton & Company, 2010), p. 320.

63. Originally published in the *Physical Review*, 1935, 47, pp. 777–780. Quoted from Kumar, pp. 304–305. The following textual quotations are also from Kumar.
64. Aspect, Alain, Philippe Grangier, and Gérard Roger, "Experimental realization of Einstein-Podolsky-Rosen-Bohm *Gedankenexperiment*: A new violation of Bell's in-inequalities," *Physical Review Letters*, 49, 1982, pp. 91–94. Quoted from Kumar, p. 350. The following quotation is also from Kumar.
65. Feynman, Richard P., *The Character of Physical Laws* (London: BBC Publications), quoted from Kumar, pp. 351–352.
66. Cf. Ne'eman, Yuval and Yoram Kish, *The particle hunters* (Cambridge: Cambridge University Press, 1983), pp. 196–222.
67. Kaku, Michio, *PARALLEL WORLDS: A JOURNEY THROUGH CREATION, HIGHER DIMENSIONS, AND THE FUTURE COSMOS* (New York: Doubleday, 2005), p. xvii.
68. Cf. Robinson, Andrew, *Einstein: A Hundred Years of Relativity* (New York: Harry N. Abrams, Incorporated, 2005), pp. 49.
69. Greene, Brian, *The Elegant Universe* (New York: W. W. Norton &Company, 1999), pp. 46–147.
70. Gerard, Piel, *The Age of Science: What Scientists Learned in the Twentieth Century* (New York: Basic Books, 2001), p. 112. The following two textual citations are also from this work.
71. Shapiro, T. Rees, "Nobel Prize-winning Physicist Survived Concentration Camp," *The Washington Post*, October 1, 1010, B7.

Index

A
Abu-I-Wafa, 15
Adelard of Bath
   renewal of scientific inquiry in the West during the 12th and 13th centuries, p.19
al-Battani, 15
Albert Magnus (Albert the Great), 26
al-Farabi, 15
al-Ghazali
   decline of Arabic science, 17, 18
   preached mystic Sufism as opposed to scientific inquiry, 17
Alhazen (Ibn al-Hatham), 15–16
al-Khwarizmi, 14, 15
   greatest Arabic mathematician, 14
al-Kindi
   Arabic Philosopher King, 14
al-Mamun, Abbasid Caliph, 14
al-Mansur, Abbasid Caliph, 14
al-Rashid, Abbasid Caliph, 14
al-Razi (Rhazes), 15
Ambrose, Bishop, 11-12
Anaxagoras, 1
Anaximander, 1

Anaximines, 1
Apollonius of Perge, 3
Aquinas, Thomas, 17, 26, 62
Arbuthnot, John, 127
Archbishop Ascanio Piccolomini, 74
Archimedes, 3, 40, 51
Archytas of Tarentum, 2
Aristarchus, 1, 33
Aristotle, 2, 20
   cosmological system, 22–23
   Metaphysics, 23
   scholastic revisions, 25, 31
Aspect, Alain, 145
Augustine, Bishop of Hippo, 11–12
Averroës, t16
Avicenna, Ibn Sina, 16

B
Babington, Humphrey, 128
Bacon, Francis, 26, 44
   Aphorisms, 85, 91, 92
   believed in the Genesis account of Adam's fall, 92
   criticisms of Aristotle, 88

divine revelation *vs.* sensory knowledge of nature, 86
flawed methodology, 87, 88, 89, 90, 92
Idols of the Cave, 85
Idols of the Market place, 86
Idols of the Theatre, 86
Idols of the Tribe, 85
life's achievements, 82, 83
emphasis on mathematics, 87
mistaken criticisms of his contemporaries, 82, 87
method of induction, 88–89
The Great Instauration, 84, 87, 92
The New Organon, 84, 87, 90, 91, 92
Warhaft, Sidney, 83, 86, 91
Bacon, Roger, 26, 27, 45
Beard, A. W., 7, 8
Behe, Michael, 134, 135
Bell, John Stewart
hidden variable theory, 144
inequality principle, 145
spin correlations, 145
Berkeley, 103
Bohm, David, 144
Bohr, Niels, 141
Copenhagen interpretation, 143, 146
Institute for Theoretical Physics, 141
nonrealist, nonlocality interpretation, 146
solar model of the atom, 141, 143
theory of complementarity, 143
Born, Max, 141
Boyle, Robert, 112, 124
Bradwardine, Thomas, 26, 29,
Brahe, Tycho, 38, 40, 42, 100
Browne, Edward, 137
Bruno, Giordano, 73
Buridan, Jean, 30–31

C

Caccini, Tommaso
denunciation of the Galileists, 57
Catholic Church during the Middle Ages, 12
Causal powers, 21
celestial realm, 22, 23
Charpak, George, 148–49
Clagett, Marshall, 25

Clauser, John, 145
Clavelin, Maurice, 29-30
Constantine, Emperor of the Holy Roman Empire, 4, 11
adoption of Christianity as the official religion, 11
transfer of the Empire to Constantinople, 11
Contextual realism, 149
Copernicus, Nicolas
De revolutionibus coelestium, 33
Commentariolus, 33, 34
Crombie, 20, 25, 26, 28, 29

D

Darwin, 3, 65, 135
Davisson, Clinton, 141
d'Abana, Pietro, 28, 33
Day, Dr. John, 43
de Broglie, Louis, 140
matter-wave concept, 141
de Duillier, Fatio, 129
Dembski, William A., 134–35
Descartes, René, 59, 62, 76, 81, 92
Beeckman, Isaac, 93
knowledge based on "*indubitable intuition*" and "*necessary connection*," 97, 98
Discourse on Method and Essays, 96, *101*
distinction between necessarily and contingently conjoined concepts, 96
distinguishing dreams from reality, 103, 104
divergence from Bacon's philosophy, 81
father of modern philosophy, 107
flaws in Descartes' reasoning, 99
"I think therefore I am," 97, 98
Meditations on First Philosophy, 101
metaphysics, 108
method of doubt, 92, 97
multifaceted genius, 92
Principles of Philosophy, 101, 105
repudiation of his previous doubts, 104
Rodis-Lewis, Geneviève, 94
Role of God, 99, 100
Rules for the Direction of the Mind, 94
vortices hypothesis, 123
Dewhust, Kenneth, 8
Dioptrice, 40
Drake, Stillman, 57, 60, 61, 71, 72, 73, 78–79

INDEX | 159

Duns Scotus, John, 28
Durant, Will, 8

E
Eddington, Arthur
   telescopic confirmation of Einstein's general theory of relativity, 139
Einstein, 101, 110, 146
   "annus *mirabilis*," 140
   Brownian motion, 110, 131
   Einstein, Rosen, and Podolsky (EPR article), 143, 144
   overthrow of Newtonian mechanics, 110
   photoelectric effect, 140
   photons, 140
   reinterpretation of Newton's gravitational framework as a curved four-dimensional space-time, 110
   revision of Newton's cosmological framework, 110, 139
   theories of relativity, 120
Empedocles' four elements, 1
Emperor Rudolph II, 38
Euclid, 2
Eudoxus of Tarentum, 2, 3, 51
Everett, Hugh, III, 146

F
Fabricius, Johann, 40
Fakhry, Majid, 14, 15, 17,
Falwell, Jerry, 138
Faraday, Michael, 45
   theory of electromagnetism, 147
Ferris, Timothy, 14
Feynman, Richard, 145
Flamsteed, John, 127
Francis, Alban, 128
Freedman, Stuart, 145

G
Galen, 3, 15
Galileo, 35, 51, 83
   appearance before Inquisition in Rome, 71
   Bellarmine, Cardinal, 55, 57, 72
   De Motu, 52
   dialogues Concerning Two New Sciences, 53, 74
   Dialogue Concerning the Two Chief World Systems, 51, 61, 62, 74
   Discourse on Bodies on or In Water, 55
   construction of the "spyglass," 53
   convicted by the Cardinals of the Inquisition, 73
   "disgrace of the century," 73
   diagrams of the surface of the moon, 5
   discovery of the rings of Saturn, 55
   Edict of 1616, 71
   epistemological dualism, 59
   explanation of projectile motion, 74
   expression of scientific skepticism, 55, 56
   freed scientific inquiry from the Catholic Church, 79
   Il Saggiatore (*The Assayer*), 58, 59
   incline plane investigations, 53
   Inquisition, 60, 61, 68–71, 73
   internment in Santa Croce in Florence, 78
   invitation from Archbishop Ascanio Piccolomini of Siena, 74
   La Bilancetta, 52
   law of accelerated free fall, 53
   Letter to Castelli distinguishing the domains of religion and science, 55, 56, 72
   "mathematics the language of nature," 60
   moons (satellites) of Jupiter, 55
   return to his villa in Arcetri where he wrote the *Dialogues Concerning Two New Sciences*, 74
   projectile motion, 74–75, 78
   Reply to Ingoli, 61, 64
   Sidereus Nuncius (*The Starry Messenger*), 53
   sun spots, 55
   trial by the Inquisition, 72
Gassendi, Pierre, 112
Gell-Mann, Murray
   "the eightfold way," 146
Germer, Lester, 141
Gilbert, William, 26, 33, 39, 41, 43, 56. 92
   De Magnete, 43, 45
   discovery that opposite poles attract, 46
   invention of the Terrella, 45
   meridians and longitudes, 47
   principles of the compass, 45
   properties of loadstones and magnets, 4

Grand Unified Theory (GUT), 127, 148
Grassi, Orazio, 58
Greeks
  intellectual tradition, 11
  mathematical and scientific legacy, 1–4
Greene, Brian, 147
Gregory, David, 129
Grosseteste, Robert, Bishop of London, 25, 60
  contributions, 26
  founded a distinguished school of scientific inquiry at Oxford University, 25
  revisions of Aristotle and later scholastics, 26

H
Hallam, Henry, 44
Harvey
  circulation of the blood, 104
Hawking, Steven, 101, 134, 149
  multiuniverse, 148
Heisenberg, Werner, 65, 141
  principle of uncertainty, 142, 143
  Scattering Matrix, 141
Heliocentric system, 35, 68
Hellenistic Period, 2
Heraclides of Pontus, 1
Hero of Alexandria, 3
Herophilus, 1
Herschel, J. F. W., 44
Higgs Boson or "God Particle," 148
Hipparchus, 3
Hippocrates of Cox, 1, 2, 3
Homocentric universe, 33
Hubble's advanced telescopes and space satellites, 46, 148
Hytesbury, William, 29

I
Ibn Sina, 16
Impetus theory, 30
Inertial motion, 31, 78
Ingoli, Francesco, 61–63

J
Jesuit College of La Flèche, 93
John of Dumbleton, 26, 29
Johnson, Phillip, 134–35

Jonson, Ben, 82
Jordan, Pascual, 141

K
Kaku, Michio
  parallel worlds, 146
Kepler, Johannes, 3, 33
  Astronomia Nova, 39, 40, 41
  celestial mechanics and clockwork universe, 41
  Dioptrice, 40
  Epitome Astronomie Copernicanæ, 41
  Harmonice Mundi, 40
  Mysterium Cosmographicum, 37, 38, 39, 52
  periodic law, p. 40
  Pythagorean geometric solids, 36, 37
  sun's gravitational force, 40
  Tabula Rudolphine, 42
  three laws of motion, 40
King James II, 128
Koestler, Arthur, 39, 41,113
Koyré, Alexander, 76
Kramers, H. A., 141
Kuhn, Thomas, 32
Kumar, Manjit, 143

L
Lemaître, George, 146
Leucippus and Democrates, 1, 124
Locke, John, 59, 125, 129
  microscopical vision, 131
Longinus
  early church father, 4
Lord Jeffreys, 128
Luther, Martin, 12, 33–34

M
Mach, Ernst, 124–25
Magnetism, 45, 46, 48, 49
Marquis de l'Hôpital, 127
Maxwell, James Clerk, 45
  electromagnetic fields of force, 147
Mersenne, 62, 76, 94
Michelson-Morley experiments, 120
Miller, Kenneth R., 135
Mlodinow, Leonard, 134

Mongols, 17
Montague, Charles, 135–36
Muhammad
   origin of the Koran, 8

N
Napier John, 42
Ne'eman, Yuval
   "the eightfold way," 146
Newton, Isaac, 83, 136
   absolute motion and rest, 120
   Æthereal medium, 133
   anni mirabilis, 112
   Barrow, Isaac, 115
   Bernoulli, Johann, 136
   Cohen, Bernard, 130, 132
   Collins, John, 115
   comparative evaluation of Newton and Einstein, 109
   corpuscular theory, 125, 131, 132
   deistic conception of God, 134
   De motu corporum (On the Motion of Bodies), 118
   $F = ma$, 121, 126
   gravitational force, 113, 122
   Halley, Edmund, 117, 118, 129
   Hooke, Robert, 114, 117
      controversy with Newton, 129
   Huygens, Christian, 117,129
   incomparable achievements, 109
   interred in Westminster Abbey,138
   Kepler's third law, 113
   laws of motion, 31, 114, 121
   light as a spectrum of colors, 114, 131
   Lord Jeffreys, 128
   Lucasian Professorship, 116
   Marquis de L'Hôpital, 127, 136
   Miller, Kenneth R, 135
   Moore, Keith, 110, 111, 113
   nature of physical reality, 115, 124
   Newton's rings, 114, 115, 131
   occult force of gravitation, 123
   Oldenburg, Henry, 117, 125
   Opticks, 129, 130
   papers on "The Lawes of Motion," 114
   political activities, 136, 137
   president of the Royal Society, 137
   Principia Mathematica
      Vol. I,110, 114, 119, 121, 122, 123–26
      Vol. II, 123, 133
   prism experiments, 131
   probing the interior of the atom with alpha particles, 130, 131
   Queries, 130, 131, 132, 133
   reflecting and refracting light, 131
   reflecting telescope, 116, 117
   rejection of absolute space and time, 139
   revision of his cosmological system, 139
   Small, Robert, 114
   studies in early church history adopting the Arian Creed over the Athanasian Creed, 116
   theory of planetary motion, 117
   Warden and Master of the Mint, 135, 136
   Westfall, Richard S., 112, 115, 116, 117, 135,137
   Whewell, William, 44

O
Ockham, William, 26
Oersted, Hans Christian, 45
Oresme, Nicole, 28, 30, 33, 74
Origen
   early Church Father, 4
Osiander, 34

P
Pauli, Wolfgang (*"infant terrible"*), 141
Peachell, John, 128
Pecham, John, 26, 28
Penrose, Roger, 147
Penzias, Arno, 147
Persinger, Michael A., 7
Philolaus, 1
Piccolomini (Archbishop Ascanio), 74
Piel, Gerard, 148
Planck, Max, 65
   constant, 142
   $E = h\nu$, 140
   explanation of blackbody radiation, 132
   introduction of quanta of energy, 140
   quantum mechanics, 83, 110, 139, 140

Planetary orbits, 38
Plato, 2, 15, 36
   Receptacle, 2
   Realm of Forms, 2
   Demiurge, 2
   Academy, 2
Plotinus, 4-9
   divine emanations, 4, 9
   Ennaeds, 4
Podolsky, Boris, 143
Pope Benedict XVI, 13
Pope Paul III, 34
Pope Paul V, 58
Pope Urban VIII, 35, 51
Porphyry, 4
Priestly, Joseph, 44
Ptolemy, 1, 3, 11, 28, 34
   Almagest, 3, 15
   astronomy, 28, 34
   geocentric system, 1
   motion, 23, 48
   scientific inquiry, 65
   theory of vision, 15–16, 25
Pythagoras, 1
   five geometric solids, 36
   polyhedra orbs, 37

Q
Quantum electrodynamics (QED), 147
Quantum mechanics, 83, 110, 139, 140

R
Recession of galaxies and inflationary universe, 147
Rectilinear motion of meteors, 63
Rheticus, 34

Ricci, Ostilio, 51, 52
Rinuccini, Francesco, 73
Robertson, Pat, 138
Robinson, Howard, 147
Robison, John, 45
Rosen, Nathan, 143
Rushd, Ibn, 16–17

S
Saint Paul, 12
Saint Peter, 12
Scheiner, Christopher (Apelles), 55
Schrödinger, Erwin, 141
   wave mechanics, 142
Singer, Charles, 2, 15, 16
Slater, E., 7, 8
Slater, John, 141
Snell, Willibrord, 101
Stukeley, William, 112
Swineshead, Richard, 29

T
Tertullian, 86
Thomson, Thomas, 44

V
Vision, 5–7, 15, 17, 28, 130

W
Weinberg, Steven, 149
Wickins, John, 129
William of Orange, 128
Wilson, Robert, 147
Witelo, 28
Wycliffe, John, 12